FEAR AT THE DISCOTHEQUE

Mr. Smith investigates the control society as a challenge, as an opportunity to observe and analyse the reciprocal link between fear and space, creating the character of Mr. Smith for the purpose. Mr. Smith, hyper-aware of his living environment, shapes his personality within an information, recording, control, capsular and anxiety environment – an environment which has repercussions on the way in which society is organized on an urban and architectural level. A condition in which disinformation, mental and physical capsules, control and data fight over life and death, truth, myth and perception. Mr. Smith aims to interpret and anticipate the (un-)conscious and (un-)intentional motivations and the mechanism behind this society of control – in which global and local initiatives struggle for hegemony. Within this condition, to Mr. Smith, response and design can only exist through a contradictory possession of the contemporary public domain. Possession, but not appreciation, nor rejection. At most interpretation and dialogue. This is how Mr. Smith tests the parameters of those who manage the control society. Figuratively and literally, Mr. Smith takes up a position in the contemporary public domain. Hyper-aware of this environment and of his own motivations, he manages, paradoxically enough, to expose the subconscious layer of our relationship between fear and space. This awareness is expressed in two design-research approaches: a techno-economic one and a socio-political one. Both approaches are not necessarily applied to arrive at solutions, but to outline and clarify dilemmas.

ANGST IN DE DISCOTHEEK

Mr. Smith onderzoekt de controlemaatschappij als een uitdaging, als een mogelijkheid om het wederzijds verband tussen angst en ruimte te observeren en te analyseren. Daarvoor riep het bureau het personage Mr. Smith in het leven. De van zijn leefomgeving hyperbewuste Mr. Smith vormt zijn karakter in een informatie-, registratie-, controle-, capsulaire en angstomgeving. Een omgeving die haar weerslag heeft op de manier waarop de maatschappij stedelijk en architecturaal wordt georganiseerd. Een conditie waarin desinformatie, mentale en fysieke capsules, controle en data een strijd leveren op leven en dood, waarheid, mythe en perceptie.
Mr. Smith beoogt de (on)bewuste en (on)bedoelde motieven en het mechanisme achter deze controlemaatschappij – waarin globale en lokale initiatieven strijden om hegemonie – te interpreteren en erop te anticiperen. Binnen deze conditie kunnen voor Mr. Smith reactie en ontwerp slechts bestaan door een tegenstrijdige aanvaarding vanhet hedendaagse publieke domein. Aanvaarding, maar geen appreciatie, noch verwerping. Hoogstens interpretatie en dialoog. Zo toetst Mr. Smith de parameters van degenen die de controlemaatschappij besturen. Mr. Smith neemt letterlijk en figuurlijk positie in het hedendaagse publieke domein. Hyperbewust van die omgeving en van de eigen motieven weet hij daardoor paradoxaal genoeg het onderbewuste van onze relatie tussen angst en ruimte bloot te leggen. Dat bewustzijn uit zich in twee ontwerp-onderzoekende benaderingen: een techno-economische en een sociaal-politieke. Beide benaderingen worden niet noodzakelijk toegepast om oplossingen aan te reiken, maar om dilemma's te schetsen en te verduidelijken.

ANGST IN DE DISCOTHEEK

FEAR AT THE DISCOTHEQUE

**PROJECT:
MR. SMITH
BIJDRAGE / CONTRIBUTION:
DR. BARBER & DR. KARANT**

PSYCHIATRIC REPORT

FEAR AT THE DISCOTHEQUE
PSYCHIATRISTS: M. VAN BEEST, D. DOEPEL, C. LINDERS, M. THEMANS AND R. WALL
THE DUTCH INSTITUTE FOR SPATIAL DISEASES
CASE: 'FEAR AT THE DISCOTHEQUE'
DOSSIER CODE: 9184
DATE: 11/12/2004

PATIENT PROFILE

PATIENT: MR. SMITH
SEX: UNSPECIFIED
AGE: 33
ORIGIN: ROTTERDAM, THE NETHERLANDS
PROFESSION: MULTIDISCIPLINARY
MENTAL DISORDER: ACUTE TECHNO-DEMOCRATIC-PHOBIA (TDP)

Dear Dr. Barber and Dr. Karant,
Below find the diagnosis of Mr. Smith.
As mentioned on the phone, TDP is not
our field of expertise. Therefore we are
pleased that you will write his prognosis,
and hereby assist us in completing the
psychiatric report.

SYNOPSIS OF PSYCHOSIS

Mr. Smith suffers from severe spatial-anx-
iety associated with techno-political and
socio-economic motives and in essence
the syndrome is structurally bipolar, expressed as both a worldwide and localized
spatial anxiety. The latter concerns the
1st Middelland Street in Rotterdam and
forms the origin of the psychosis. The two
spatial extremities of this anxiety are highly
correlated, in which Smith believes that the
1st Middelland Street serves as a proto-
typical 'fearspace', reflecting the ongoing
tension between what he perceives as the
underworld and the control-society. Smith's
fearspace is catalyzed on the one hand by
observations concerning increasing social
and urban deterioration in Rotterdam. It
is believed by Smith that these local conditions are significantly determined by
global, socio-economic inequalities, which

Geachte dokter Barber en dokter Karant,
Hierbij sturen wij u, zoals telefonisch afgesproken,
onze diagnose van Mr. Smith. Zoals we al aan-
gaven, valt TDF niet binnen ons specialisme.
Het verheugt ons derhalve dat u zich bereid heeft
getoond de prognose uit te werken en ons aldus
te helpen dit psychiatrisch rapport te voltooien.

SAMENVATTING VAN DE PSYCHOSE

Mr. Smith heeft last van ernstige ruimtevrees die
bepaald wordt door technisch-politieke en sociaal-
economische factoren. In wezen betreft het hier
een structureel bipolair syndroom dat zich uit in
ruimtelijke angsten die zowel een globaal (mondiaal)
als een lokaal karakter kunnen hebben. Die laatste
variant heeft betrekking op de 1e Middellandstraat
in Rotterdam en vormt de oorsprong van de psychose. De twee ruimtelijke extremen van deze
angststoornis zijn nauw met elkaar verbonden:
Smith is ervan overtuigd dat de 1e Middellandstraat
de prototypische 'angstruimte' vertegenwoordigt en
de aanhoudende spanning weerspiegelt tussen de
controlemaatschappij en wat hij als onderwereld
beschouwt. Smith's angstruimte wordt aan de ene
kant gekatalyseerd door zijn waarnemingen van het
verslechterende sociale en stedelijke klimaat in
Rotterdam. Smith gelooft dat deze lokale omstandigheden in belangrijke mate bepaald worden door
wereldwijde sociaal-economische ongelijkheden,

PSYCHIATRISCH RAPPORT

ANGST IN DE DISCOTHEEK
PSYCHIATERS: M. VAN BEEST, D. DOEPEL, C. LINDERS, M. THEMANS EN R. WALL
HET NEDERLANDS INSTITUUT VOOR RUIMTELIJKE AANDOENINGEN
GEVAL: 'ANGST IN DE DISCOTHEEK'
DOSSIERCODE: 9184
DATUM: 11/12/2004

PATIËNTENPROFIEL

PATIËNT: MR. SMITH
GESLACHT: NIET NADER OMSCHREVEN
LEEFTIJD: 33
OORSPRONG: ROTTERDAM, NEDERLAND
BEROEP: MULTIDISCIPLINAIR
PSYCHISCHE STOORNIS: ACUTE TECHNO-DEMOCRATISCHE FOBIE (TDF)

in turn contribute highly to the proliferation of crime and the threat of terrorist attack. This leads to a definition of Smith's first fear – 'underworld' (i.e. anti-globalization activists to crime/terrorist syndicates). On the other hand, Smith observes that public and private institutions counter underworld activities, by means of enforced 'control mechanisms', such as surveillance technologies. According to Smith, the 1st Middelland Street is coined the 'surveillance route' by the Rotterdam Police Department, who constantly monitor and control the area. This leads to a definition of Smith's second fear - 'control society' (i.e. city, national and international institutions).
The patient believes that current interventions by the control-society to combat underworld activity in the 1st Middelland Street, or any city or public space for that matter, are in vain. He fears that enforced surveillance technologies, policing and preventive urban planning will only lead to short-term, inefficient solutions. This forms the epicenter of Smith's syndrome – 'that the world in general is blind to structural change, due to greed, ignorance and negligence'. Smith has compulsive feelings of responsibility to alter 'our' multidimensional fearspace, but often pessimistically

die op hun beurt een grote rol spelen in de verbreiding van criminaliteit en de dreiging van terrorisme. Dit leidt tot een omschrijving van Smith's eerste angst: de onderwereld (van antiglobalistische activisten tot misdaad- en terreursyndicaten). Aan de andere kant ziet Mr. Smith dat publieke en private instellingen de activiteiten van de onderwereld tegengaan met behulp van verscherpte 'controlemechanismen' zoals bewakings- en beveiligingstechnologieën. Volgens Smith wordt de 1e Middellandstraat door de Rotterdamse politie, die de hele buurt voortdurend in het oog houdt, ook wel de 'surveillanceroute' genoemd. Dit leidt tot een typering van Smith's tweede angst: de controlemaatschappij (controle door instellingen op stedelijk, nationaal en internationaal niveau). De patiënt is van mening dat de huidige ingrepen van de controlemaatschappij, gericht op de bestrijding van onderwereldactiviteiten in de 1e Middellandstraat, of waar dan ook in een stad of de publieke ruimte, zinloos zijn. Hij vreest dat de huidige beveiligingstechnologieën, nadrukkelijk politieoptreden en preventieve stedenbouwmethoden, slechts leiden tot ondoelmatige oplossingen voor de korte termijn. Hier ligt het epicentrum van Smith's syndroom: 'dat de wereld in het algemeen als gevolg van hebzucht, onwetendheid en onverschilligheid blind is voor structurele veranderingen'. Mr. Smith heeft last van een dwangmatig verantwoordelijkheidsgevoel; aan de ene kant voelt hij de drang 'onze' multidemensionale angstruimte te

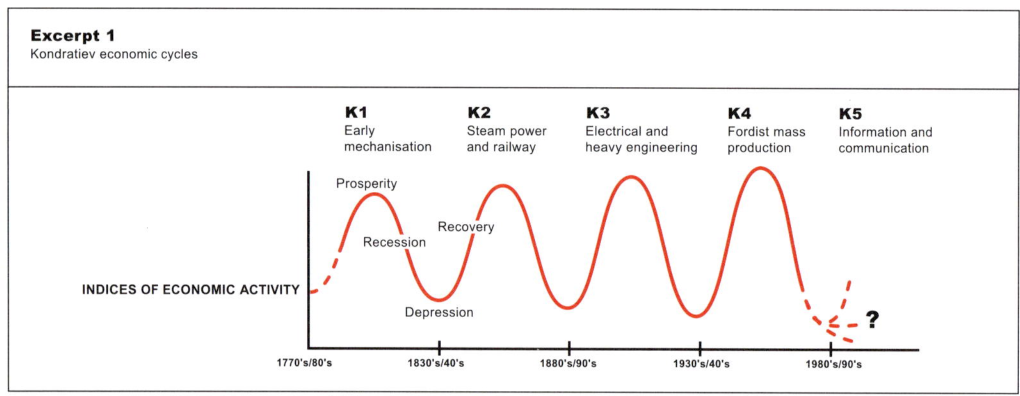

"THE NOTION THAT ECONOMIC GROWTH OCCURS IN A SERIES OF LONG-WAVES 'KONDRATIEV' WAVES OF MORE OR LESS 50 YEARS' DURATION GOES BACK TO THE 1920S. IN THE GRAPH FOUR COMPLETE K-WAVES ARE IDENTIFIED; WE ARE NOW IN THE MIDST OF A FIFTH. EACH WAVE MAY BE DIVIDED INTO FOUR FAZES: PROSPERITY, RECESSION, DEPRESSION AND RECOVERY. EACH WAVE TENDS TO BE ASSOCIATED WITH PARTICULAR SIGNIFICANT TECHNOLOGICAL CHANGES. THE FIFTH K-WAVE IS ASSOCIATED PRIMARILY WITH INFORMATION TECHNOLOGY."
(DICKEN, 2003)

withdraws into emotions of lethargy. Smith's personality consists of four alter egos: the implicit (agonistic), the explicit (conflict), the self (territorial) and the other Mr. Smith (interactive).
Furthermore, Smith is convinced that the system that the control society is protecting is in fact the root problem, and explains, 'we should not be maintaining this system, but should instead transform it'. Smith wonders whether the pursuit of a universally democratic system, which facilitates the values of all societies, will effectively reduce or transform fearspace between and within cities.
Symptoms of Smith's psychosis are:
(1) has conceived integral, vivid, multi-dimensional fearspaces; (2) observes the world through techno-political and socio-economic filters; (3) is suspect of the contemporary democratic system; (4) has acute delusions of social responsibility; (5) is a specialist in lateral relations, focusing on cause and effect; (6) is obsessed with intervening in the system; (7) symptomatically fluctuates between lethargy and over-optimism; (8) peculiarly substantiates psychotherapy using personal research and visualizations; (9) intends to transmit his psychosis to a general public – in the

veranderen, maar aan de andere kant trekt hij zich vaak terug in een pessimistische lethargie. Smith's persoonlijkheid is samengesteld uit vier alter ego's: de impliciete (strijdlustige), de expliciete (conflictueuze), de eigen (territoriale) en de andere (interactieve) Mr. Smith.
Bovendien is Mr. Smith ervan overtuigd dat het systeem dat door de controlemaatschappij wordt beschermd in feite juist de kern van het probleem vormt. 'In plaats van het in stand te houden, zouden we dit systeem moeten omvormen', zo legt hij uit. Mr. Smith vraagt zich af of het streven naar een universeel democratisch systeem, dat recht doet aan de waarden van alle maatschappijen, wel een effectief middel is om de angstruimtes binnen en tussen steden te verkleinen of om te vormen.
De symptomen van Smith's psychose zijn: 1) hij heeft voorstellingen van integrale, levendige multi-dimensionale angstruimtes; 2) bekijkt de wereld via technisch-politieke en sociaal-economische filters; 3) wantrouwt het hedendaagse democratische systeem; 4) heeft acute wanen van maatschappelijke verantwoordelijkheid; 5) is een specialist op het gebied van laterale relaties, gericht op oorzaak en gevolg; 6) is geobsedeerd door interventies in het systeem; 7) fluctueert symptomatisch tussen lethargie en overdreven optimisme; 8) geeft op zonderlinge wijze vorm aan psychotherapie, met behulp van persoonlijk onderzoek en visualisaties; 9) is van plan zijn psychose over te dragen op een algemeen

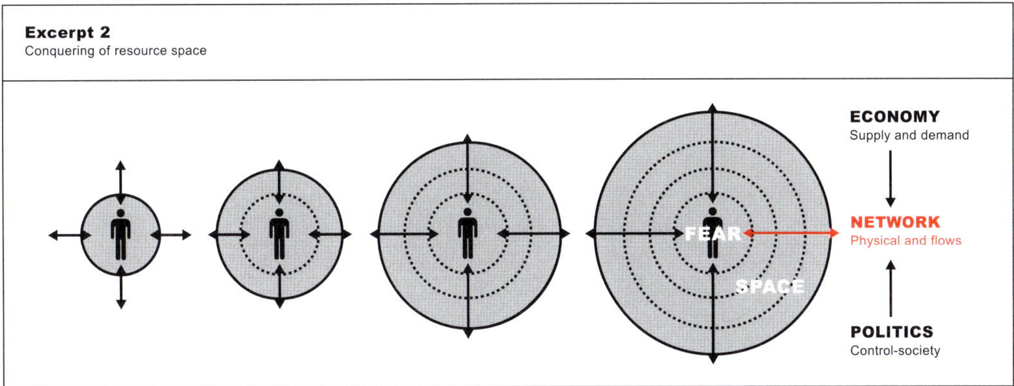

"BY GRADUALLY BECOMING A MORE INDEPENDENT VARIABLE IN THE DEVELOPMENTAL EQUATION, TECHNOLOGY REINFORCED THE CITIES ROLE AS A CENTRE OF THE CREATION AND DIFFUSION OF TECHNOLOGICAL INNOVATIONS." (BAIROCH, 1988)

form of a VJ presentation called Fear at the Discotheque.

In this report, Smith's fearspace is diagnosed in three parts: (A) Context of Patients Psychosis; (B) Patient's Filtering Mechanism and (C) Patient's Conceptual Model of Fearspace. The diagnosis is based on our observations, with detailed reference to Smith's own therapeutic research and visual excerpts. The substantial research references to the following diagnosis can be found at www.groepsportretten.nl/smith.

DIAGNOSIS
A. CONTEXT OF PATIENT'S PSYCHOSIS

In the theoretical model by the economist Kondratiev[1] [excerpt 1], Smith's fascination with socio-economic development by means of technological progress is revealed[2]. Related to this, Smith indicates in a series of steps [excerpt 2] that today's global urban network[3] is the spatio-temporal result of humanity's victory over resource space by means of city development and technological progress[4]. Smith shows that over the centuries our delimitation of urban space, as 'that place' within which we feel 'safer than outside', has incrementally

publiek – in de vorm van een VJ-presentatie met de titel 'Angst in de discotheek'.

In dit rapport is de diagnose van Smith's 'angstruimte' in drie delen opgesplitst: A) de context van de psychose, B) het filtermechanisme en C) Smith's conceptuele model van de angstruimte. De diagnose is gebaseerd op onze observaties, met uitvoerige verwijzingen naar Smith's eigen therapeutische onderzoek en visuele excerpts. De referenties met betrekking tot het onderzoek van onderstaande diagnose, kunnen worden gevonden op: www.groepsportretten.nl/smith.

DIAGNOSE
A. CONTEXT VAN DE PATIËNTS PSYCHOSE

In de weergave van een theoretisch model van de econoom Kondratiev[1] [excerpt 1], komt duidelijk Smith's fascinatie voor sociaal-economische ontwikkeling door middel van technologische vooruitgang aan het licht[2]. In samenhang hiermee probeert Mr. Smith in een reeks stappen [excerpt 2] aan te tonen dat het hedendaagse mondiale stedelijke netwerk[3] het tijdruimtelijke resultaat is van de overwinning van de mens op de bronnen en middelen via stadsontwikkeling en technologische vooruitgang[4]. Mr. Smith toont aan dat de afbakening van de stedelijke ruimte, als de plaats waar we ons veiliger voelen dan erbuiten, in de loop der eeuwen steeds verder is

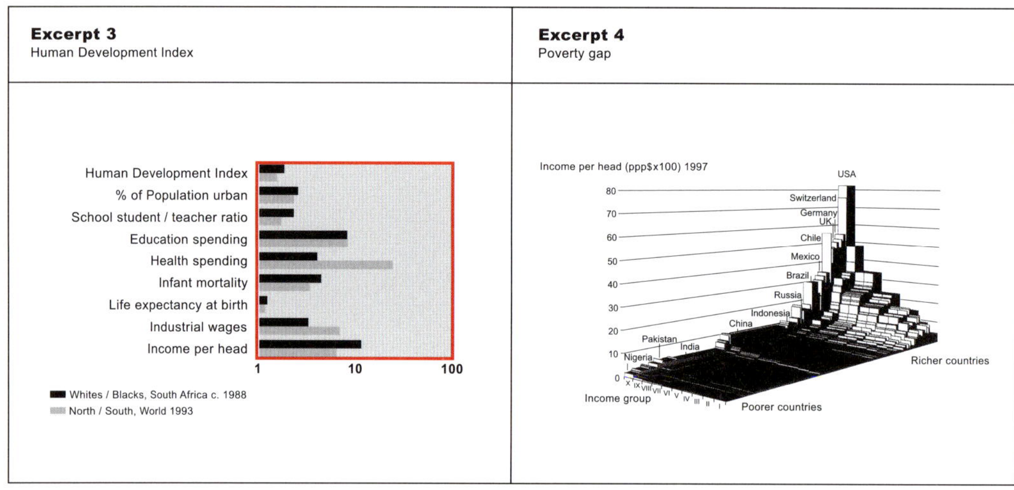

expanded from fortified villages, to cities, to our current urban networks[5]. Smith feels that as 'tool making creatures' our technology has progressively colonized real and mental space, with the false promise of an expanding domain of safety[6]. But this expensive safety system[7], only maintains the minority's status. 'Spurred by the tyranny of scarcity, we have by means of technology overcome the tyranny of distance', where Smith questions this drive to secular status[8], which is mainly to the distorted benefit of the lucky few[9]. Smith fears that global apartheid[10] has led to widespread economic disparity[11] [excerpts 3 & 4]. He ponders how our ongoing mediated awareness of inequality[12] still does not lead to economic reforms[13] emphasizing our moral insolvency[14]. How can we, Smith wonders, pursue the globalization of democracy[15], which currently only suits enduring material voracity[16] and narcissistic urban development[17]? This is promoted by false economic incentives[18], that lead to intolerable economic[19] and technological[20] disparity, such as in Africa[21], but is equally visible in the 1st Middelland Street. Besides Smith's empathy for the Third World[22], and his anticipation of an innovative trading system[23], concern is also expressed for environmental degradation[24], although Smith acknowledges the vagueness of the topic[25]. Accordingly, our hungry global economy[26] does not reveal the costs to the environment[27] and Smith is distressed about the possibility of the developing world reaching First World

opgeschoven, van versterkte dorpen tot steden tot onze huidige stedelijke netwerken[5]. Mr. Smith is van mening dat wij, als 'instrument makende' wezens, met onze technologie de werkelijke en de geestelijke ruimte steeds verder hebben gekoloniseerd, met de valse belofte het 'veilige domein'[6] nog verder uit te breiden. Maar dit kostbare veiligheidssysteem[7] dient slechts om de status van een minderheid in stand te houden. 'Aangespoord door de tirannie van de schaarste zijn we door middel van technologie de tirannie van de afstand te boven gekomen', zegt Smith, die vraagtekens plaatst bij deze drang naar een seculiere status[8], die op verwrongen wijze vooral ten goede komt aan een kleine elite[9]. Mr. Smith vreest dat mondiale apartheid[10] heeft geleid tot een wijdverbreide economische ongelijkheid[11] [excerpts 3 en 4]. Smith overdenkt de vraag waarom ons door de media vergrote bewustzijn van economische ongelijkheid[12] niet leidt tot economische hervormingen[13] en benadrukt hiermee ons ethisch onvermogen[14]. Hoe kunnen we, zo vraagt Mr. Smith zich af, blijven streven naar mondialisering van onze democratie[15], terwijl die momenteel slechts aanhoudende materiële vraatzucht[16] en een narcistische stadsontwikkeling[17] in de hand werkt? Dit proces wordt versterkt door valse economische stimulansen[18] die leiden tot ontoelaatbare economische[19] en technologische[20] ongelijkheid, bijvoorbeeld in Afrika[21], maar ook in de 1e Middellandstraat. Mr. Smith heeft niet alleen oog voor de derde wereld[22], en de noodzaak voor op de ontwikkeling van een innovatief handelssysteem[23], maar geeft ook blijk van grote zorgen om de achteruitgang van het milieu[24], al erkent hij dat het hier om een vage kwestie gaat[25]. Onze hongerige mondiale economie[26] verhult de kosten voor het milieu[27]. Mr. Smith is verontrust over de mogelijkheid dat de ontwikkelingslanden het niveau van de eerste

> **"THE INTEREST OF THE COUNTER POSITION SHOWN IN THIS DIAGRAM IS THAT IT SHOWS THAT THE DEGREE NORTH/SOUTH INEQUALITY IN THE WORLD IS STRIKINGLY SIMILAR TO WHITE/ BLACK INEQUALITY UNDER APARTHEID. THE WORLD SEEMS TO BE A MACROCOSM OF A DISCRIMINATORY SYSTEM WHICH THAT SAME WORLD ALMOST UNANIMOUSLY REJECTED AS MORALLY UNACCEPTABLE."** (SUTCLIFFE, 2001)

standards[28], for instance China's oil[29] or paper demand[30].
Smith is anxious about the direction of the current globalization project[31], thein centives of political leaders[32], such as G. W. Bush[33], the effects upon the lives of others[34] and how they are deceived by politicians[35]. Smith worries that the non-ethical stance of rich nations[36], especially America[37], will become our own problem[38] and weakness[39] and propagate the further demise of civil society[40]. 'The world is negligent[41]', in Smith's opinion, 'promoting a deficient[42] and indeterminate[43] democracy, which is rapidly being confronted by the presence of the "stranger"[44], carrying pressing new demands[45]. If these demands are ignored, a breeding ground for terrorism and crime is established, which cannot be curtailed by enforcing militant technological power[46]', leading to the further polarization of anxiety between the underworld[47] and control-society[48]. Smith feels that only by inventing new forms of liberty[49] and democracy[50], in which different[51] global citizens[52] and political institutions[53] can flourish, will our inter-meshed glocal fearspace be transformed[54].

B. PATIENT'S FILTERING MECHANISM

This paragraph focuses on the patient's peculiar way of observing reality at the local level, but before proceeding with this, four archetypical fixations of Smith are to be explained. The first [excerpt 5] signifies (a) the asymmetrical international trade

wereld[28] bereiken, en bijvoorbeeld de vraag naar olie[29] en papier[30] in China een westers niveau bereikt. Smith maakt zich zorgen over de richting van het huidige mondialiseringsproject[31], over de motivatie van politieke leiders[32] zoals G. W. Bush[33], over de effecten op de levens van anderen[34] en hoe ze worden misleid door politici.[35] Smith is bang dat de onethische houding van de rijke landen[36], met name de Verenigde Staten[37], zal uitgroeien tot ons eigen probleem[38] en zwakte[39] en zal leiden tot de ondergang van de rechtsstaat[40]. 'De wereld is nalatig',[41] meent Mr. Smith. 'Ze propageert een tekortschietende[42] en onbestemde[43] democratie die in hoog tempo wordt geconfronteerd met de aanwezigheid van de "vreemdeling[44]", die dringende nieuwe eisen[45] met zich meebrengt. Als deze eisen worden genegeerd, ontstaat er een broedplaats voor terrorisme en misdaad, die met militair-technologische machtsmiddelen[46] niet is tegen te houden.' Dit leidt tot een nog sterkere op angst gebaseerde polarisatie tussen de onderwereld[47] en de controle-maatschappij[48]. Smith denkt dat alleen door het uitvinden van nieuwe vormen van vrijheid[49] en democratie[50], waarbinnen uiteenlopende[51] bevolkings-groepen[52] en politieke instituties[53] tot bloei kunnen komen, onze lokaal en mondiaal verweven glokale angstruimte kunnen worden getransformeerd[54].

B. HET FILTERMECHANISME VAN DE PATIËNT

In deze paragraaf wordt nader ingegaan op de zonderlinge wijze waarop de patiënt de werkelijkheid op lokaal niveau bekijkt, maar daarvoor moeten eerst vier archetypische fixaties van Mr. Smith nader worden verklaard. De eerste [excerpt 5] wijst op (a) het asymmetrische internationale handelssysteem,

Excerpt 5
Archetypical fixations

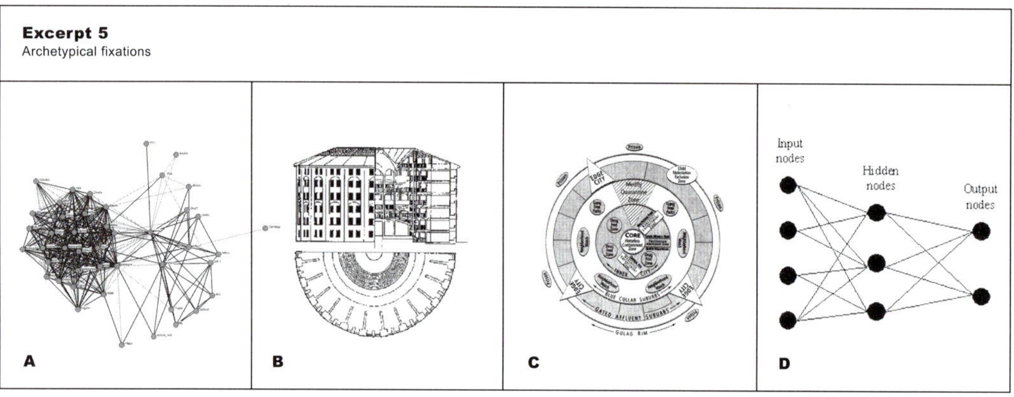

Excerpt 6
Techno-filter

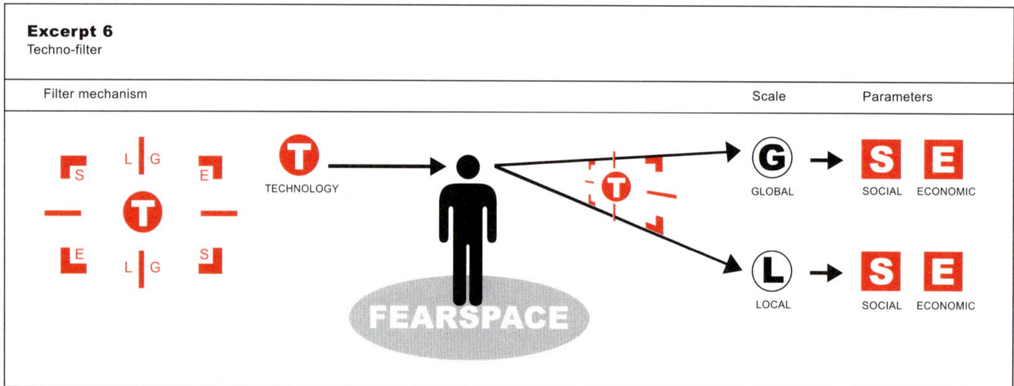

system, from which Smith believes socio-economic injustice is spawned. A system maintained by the contemporary 'control-society' and symbolized by (b) Smith's second fixation, Bentham's panopticon[55]. From this a modified anxiety – the 'super-panopticon[56]', or ultimately, the 'hyper-surveillance society[57]' has been concocted. A distorted technological system[58], forming the pivot between underworld and control-society[59], in which Smith feels we are all participating[60], by means of control activities[61] and technology[62]. From this causal duality, Smith implies how global and local fearspaces (c) are effectuated[63], by means of a techno-neurological (d) network. This leads to an intrusion of our privacy[64], the establishment of discriminatory space[65] and the generation of a risk society[66]. Based on the four archetypal fixations, Smith's fearspace is reduced to an abstract diagram (techno-filter), which demonstrates his everyday spatial perception [excerpt 6]. Hereby we illustrate a psychosis in which any daily observation made by Smith is

dat volgens Smith aan de bron ligt van de sociaal-economische ongelijkheid. Een systeem dat door de hedendaagse 'controlemaatschappij' in stand wordt gehouden en gesymboliseerd wordt door (b) Smith's tweede fixatie, namelijk Jeremy Benthams panopticum[55] (koepelgevangenis). Hieruit is een gemodificeerde angst voortgekomen – het 'super-panopticum[56]' en uiteindelijk de 'hypersurveillance-maatschappij[57]': een vervormd technologisch systeem[58] dat de spil vormt tussen de onderwereld en de controlemaatschappij[59]. Volgens Mr. Smith zijn wij allemaal bij dit systeem betrokken[60], via controle-activiteiten[61] en technologie[62]. Smith suggereert dat door deze oorzakelijke duali-teit, mondiale en lokale angstruimtes (c) tot stand worden gebracht[63], via een techno-neuraal (d) netwerk. Dit leidt tot een inbreuk op onze privacy[64], het ontstaan van discriminatoire ruimtes[65] en de totstandkoming van een risicomaatschappij[66]. Smith's angstruimte, die gebaseerd is op deze vier fixaties, kan worden teruggebracht tot een abstract schema (technofilter), dat zijn dagelijkse perceptie van de ruimte laat zien [excerpt 6]. Het illustreert een psychose waarbij alle dagelijkse observaties van Mr. Smith tegelijkertijd worden beleefd als

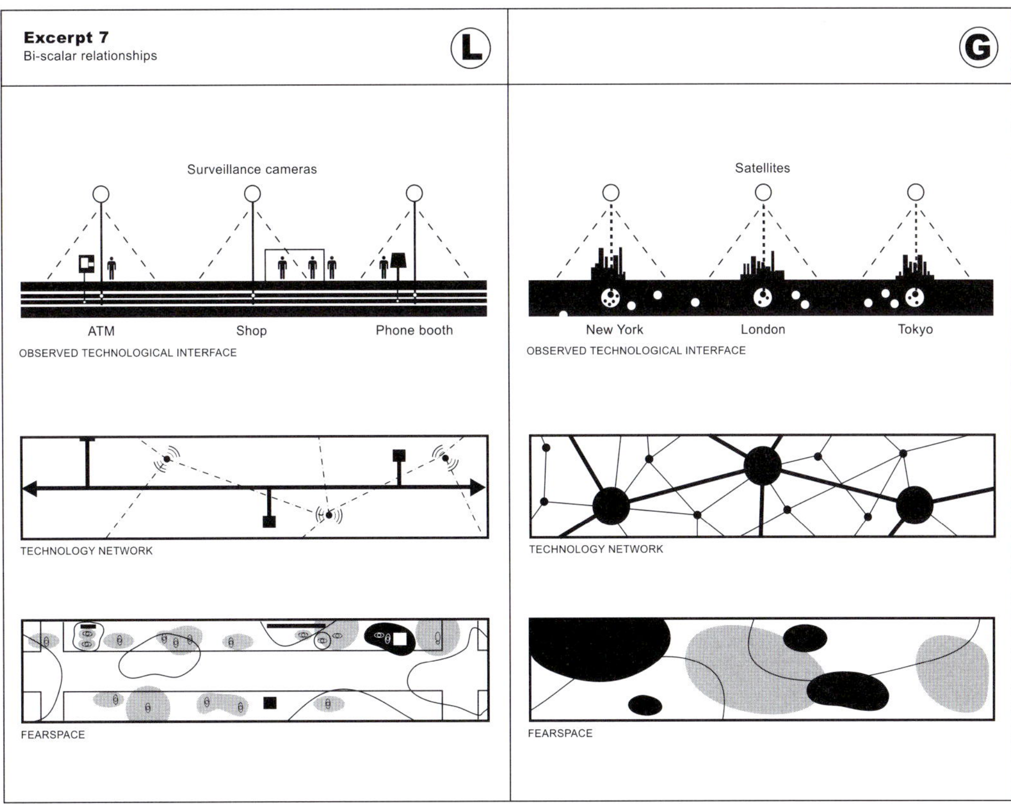

simultaneously perceived as a local and global anxiety, and is further subdivided into social and economic categories. He cannot distinguish between scales, and instead sees them as integrally connected. Smith's fearspace is therefore glocal. In [excerpt 7], we see how the patient relates for instance a surveillance camera[67] at the local scale, with satellite surveillance at the global scale. These bi-scalar relationships are divided into observed technological interface, subsequent network and the consequential fearspace, which in turn represents the tension between control-society and underworld. Underlying this tension, we find Smith's moral dilemma, concerning inequality and the desire for true democracy.

In [excerpt 8a-d], the exact trajectory of Smith's local fearspace is shown. It depicts the 1st Middelland Street, where the patient's psychosis originated and although the images are self-explanatory, it is important to note how Smith evaluates reality.

lokale en mondiale angsten die verder worden opgesplitst in maatschappelijke en economische categorieën. Hij kan geen onderscheid maken tussen de verschillende schaalniveaus, maar ziet ze als integraal met elkaar verbonden. Smith's angstruimte is dus 'glokaal'. In [excerpt 7] zien we bijvoorbeeld hoe de patiënt een bewakingscamera[67] op lokaal niveau verbindt met het systeem van observatiesatellieten op mondiale schaal. Deze biscalaire verbanden worden verdeeld in een geobserveerde technologische interface, vervolgens een netwerk en uiteindelijk een angstruimte, die op haar beurt het spanningsveld weerspiegelt tussen de controlemaatschappij en de onderwereld. Ten grondslag aan dit spanningsveld ligt Smith's morele dilemma, dat te maken heeft met ongelijkheid en het verlangen naar ware democratie.
In [excerpt 8a-d] wordt het volledige traject van Smith's lokale angstruimte getoond. Afgebeeld is de 1e Middellandstraat, waar de psychose van de patiënt zijn oorsprong vindt. Hoewel de beelden voor zichzelf spreken, is het belangrijk op te merken hoe Mr. Smith de werkelijkheid evalueert.

Fear at the Discotheque   Mr. Smith

Surveillance route
Rotterdam - Middelland

Today public space is unimaginable without surveillance cameras. These are installed to, objectively, reduce the number of crimes (preventative), and subjectively, to increase the feeling of safety (repressive).

1e Middellandstraat, West Kruiskade, Kruisplein, Stadhuisplein and Central Station. Control room at Police Station Eendrachtsplein.

There are a total of 82 surveillance cameras in six areas in Rotterdam (twice as many as in 2002). A 280.000 euro investment.

Excerpt 8a
Reconstruction of the Middelland Street fearspace

31/08/2004   13:07:11 (GMT + 02:00)   MIDDELLAND

Camera surveillance is expected to become the fifth utility, alongside electricity, water, gas and telephone.

DEPENDENCY

Surveillance camera

Today when we walk the streets and are not sure if a surveillance camera is watching us or not, we are acting to the same panoptic logic as then. [I refer to the British philosopher Jeremy Bentham, the founder of the doctrine of Utilitarianism who began working on a plan for a model prison called the 'Panopticon' in 1785.]

UNCERTAINTY

VULNERABILITY

It is not only real problems that create insecurity and feelings of uncertainty, but also the less tangible ones, like the vulnerability of modern society to calamities, and the threats inherent in the advance of technology and science. Contemporary wealth and the progress of science have increased our safety and security, but cannot prevent citizens from these unpredictable risks.

That our lives have become embedded in technology-dependent surveillance practices and that it is this system that produces amoral, self-referential effects, is unfortunately not in question.

We live in a high-risk society of interdependent complex risks, instead of personal and controllable risks.

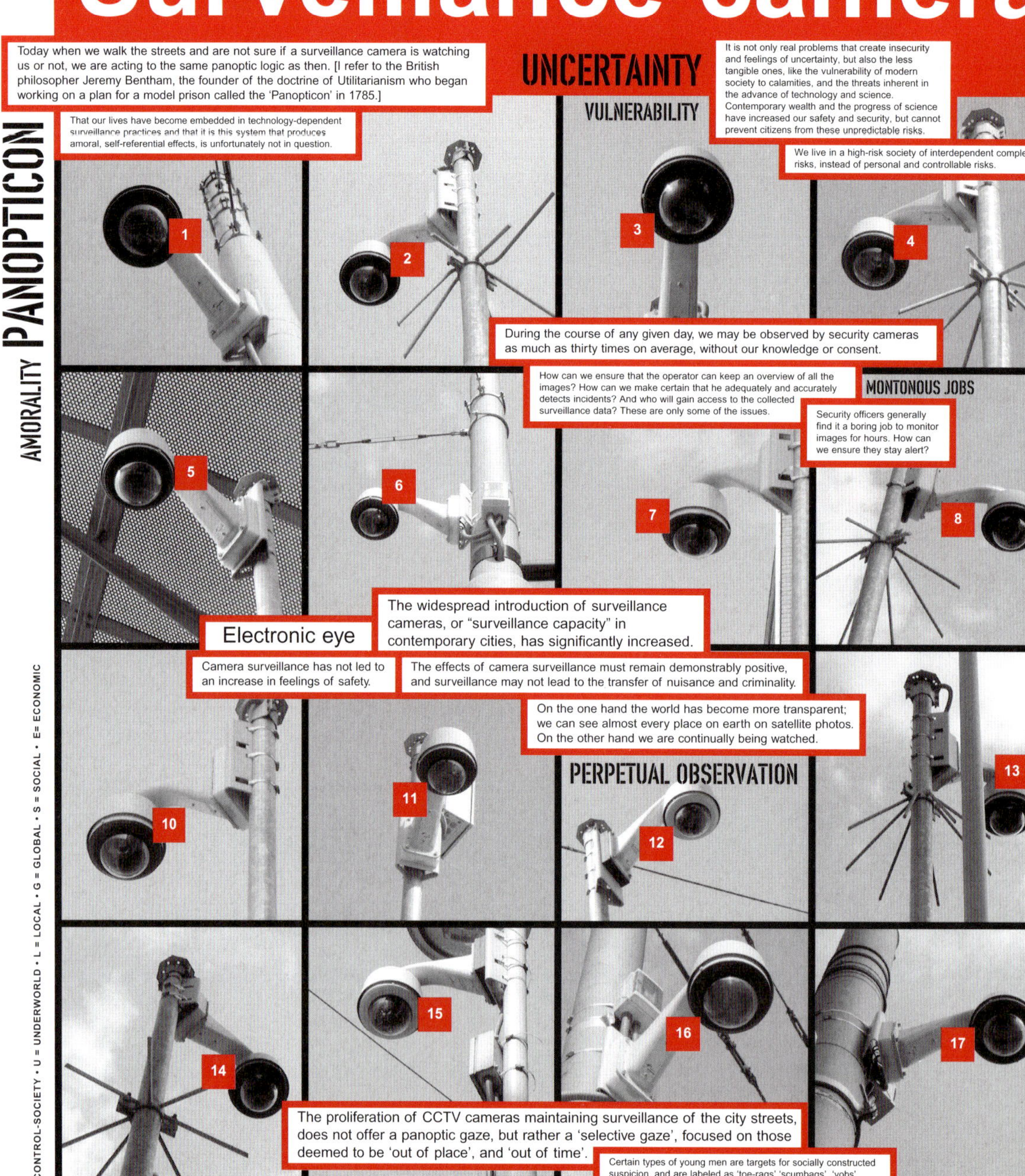

During the course of any given day, we may be observed by security cameras as much as thirty times on average, without our knowledge or consent.

How can we ensure that the operator can keep an overview of all the images? How can we make certain that he adequately and accurately detects incidents? And who will gain access to the collected surveillance data? These are only some of the issues.

MONTONOUS JOBS

Security officers generally find it a boring job to monitor images for hours. How can we ensure they stay alert?

The widespread introduction of surveillance cameras, or "surveillance capacity" in contemporary cities, has significantly increased.

Electronic eye

Camera surveillance has not led to an increase in feelings of safety.

The effects of camera surveillance must remain demonstrably positive, and surveillance may not lead to the transfer of nuisance and criminality.

On the one hand the world has become more transparent; we can see almost every place on earth on satellite photos. On the other hand we are continually being watched.

PERPETUAL OBSERVATION

The proliferation of CCTV cameras maintaining surveillance of the city streets, does not offer a panoptic gaze, but rather a 'selective gaze', focused on those deemed to be 'out of place', and 'out of time'.

Certain types of young men are targets for socially constructed suspicion, and are labeled as 'toe-rags' 'scumbags', 'yobs', 'homeless low-life', 'drug-dealing scrotes' and 'big issue scum'.

In the UK, black people in particular, 'are between one and a half and two and a half times' more likely to be under surveillance.

AMORALITY PANOPTICON

C = CONTROL-SOCIETY • U = UNDERWORLD • L = LOCAL • G = GLOBAL • S = SOCIAL • E = ECONOMIC

EXCLUSION   RACISM

LOSS OF PRIVACY
CURTAILMENT OF RIGHTS

Angst in de discotheek   Mr. Smith

50% of the Dutch population believe that the recent expansion of the European Union will lead to an increase in crime.

The police respond to approximately four million incidents per annum, twenty-five percent of which involve theft, vandalism and violence, or are traffic incidents.

Two thirds of the Dutch population believe it would reduce crime if all citizens were registered in a National Crime Prevention database.

Seven out of ten people believe that the threat of severe punishment is the only way to deter criminals.

Approximately half the Dutch population is in doubt as to whether police intervention is effective in crime prevention and reduction.

Safety versus privacy

Extreme regulations that limit our privacy induce feelings of insecurity and anxiety.

INSECURITY

2001 there have been more surveillance on the streets over weekends.

Street watch

Increasing civilian responsibility

Communities and corporate organizations are increasingly expected to take responsibility for their own safety and security.

Central Station

26 27 28 29 30 31 25

BREACH OF PRIVACY

Central Station is a safety risk area, giving police the right to carry out preventative searches. [Twenty-four cameras.]

60% of violence around CS occurs between midnight and eight AM on Saturday and Sunday.

EXCLUSION

Middelland Street

15 16 17 12 10 13 14 08 11 05 07 06 03 04 09 01 02

LOSS OF CHOICE

A restrictive foreign policy, restrictions on and privatisation of social security, signal the emergence of a government that demands closer monitoring of its citizens.

Checklist:
1. Construct a written argument that underlines the importance of camera surveillance.
2. Substantiate why public and corporate concerns, weigh more heavily than the privacy of citizens.
3. Explain why property cannot be protected in any way other than with the use of surveillance cameras.
4. Camera surveillance should form part of a set of measures that maintain order in the public domain.
5. Camera surveillance must be selectively implemented and only vital information recorded.
6. It must be made obvious that surveillance cameras are operating in any given location. Failure to do so should result in penalties.
7. Covert surveillance in public spaces (of which the public is unaware) will soon be punishable by law. This has already been implemented in shops and cafes.
8. Define in a protocol how individuals can obtain information concerning surveillance recordings of themselves and their civil rights in this regard.
9. Video material may not be kept longer than necessary. A suggested guideline is up to seven days, with the exception of recordings containing evidence of a misdemeanor.
10. The CBP has the authority to fine individuals who fail to code images correctly, or do not report an incident in a timely manner.

PARANOID SOCIETY

Violation of privacy is not the only concern, so is the nature of the society that is slowly but surely being created. A paranoid society based on untrustworthiness, in which everyone is a potential suspect, unless the computer says otherwise.

Surveillance images are monitored live, twenty-four hours a day, seven days a week. In addition, the police can access them at any given time. The privacy regulations of the Council for Protecting Citizen Data are applicable. These recommend that data be stored for a minimum of 72 hours. If within the agreed term, no reason is found to justify saving them any longer, they are destroyed.

cameratoezicht

In public spaces and roads everyone is visible. Not just the fact that you are there, but also how you behave. Citizens have only a partial amount of privacy, and this decreases rapidly since the introduction of more cameras.

The placement of cameras excludes the filming of non-public spaces, and often a part of the cameras field of vision is restricted for this purpose.

CCTV

Do-it-Yourself

Spycam
Audio Spycam
Wireless Spycam
Pencam
Microcam
Audio Microcam
Stealthcam
ll-Video
m
Processo
y Cam
Mini Security Cam
Day/Night Cam
Outdoor Cam
Weatherproof Cam
Guardian Angel Babycam
Guardian Angel Handheld
Guardian Angel TV
Guardian Angel Wireless Camera
Securacam
Securaview Monitor Kit
Wireless Monitor Kit
Wireless Camera
Quad Monitoring Kit
Alert2 Plus
Alert4
Bulletcam
C-series Camera
Audio Colorcam
D-series Camera
Domecam
Detector Cam
Imitation Smoke Detector Cam
Imitation Security Cameras
Imitation Domecam
Simulated Security Camera

324 Security companies are registered in the Yellow Pages.

There are currently 30 000 Dutch workers employed in the security industry.

In 1999, the turnover of Dutch security companies was 719 million euro. The annual growth rate in this sector is fourteen percent.

TERRITORY

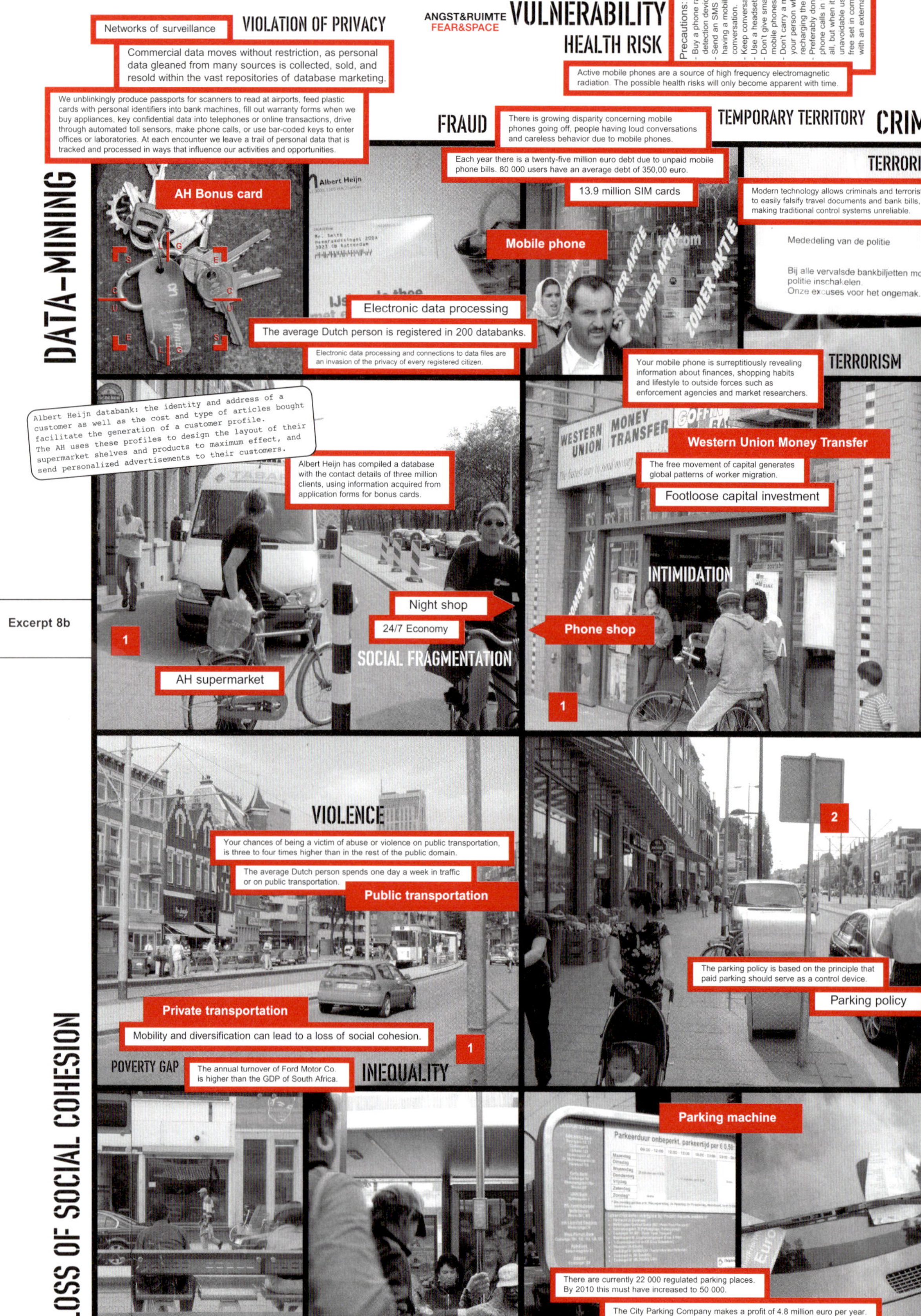

VIOLATION OF PRIVACY

Networks of surveillance

Commercial data moves without restriction, as personal data gleaned from many sources is collected, sold, and resold within the vast repositories of database marketing.

We unblinkingly produce passports for scanners to read at airports, feed plastic cards with personal identifiers into bank machines, fill out warranty forms when we buy appliances, key confidential data into telephones or online transactions, drive through automated toll sensors, make phone calls, or use bar-coded keys to enter offices or laboratories. At each encounter we leave a trail of personal data that is tracked and processed in ways that influence our activities and opportunities.

DATA-MINING

AH Bonus card

Albert Heijn databank: the identity and address of a customer as well as the cost and type of articles bought facilitate the generation of a customer profile. The AH uses these profiles to design the layout of their supermarket shelves and products to maximum effect, and send personalized advertisements to their customers.

Albert Heijn has compiled a database with the contact details of three million clients, using information acquired from application forms for bonus cards.

Electronic data processing

The average Dutch person is registered in 200 databanks.

Electronic data processing and connections to data files are an invasion of the privacy of every registered citizen.

Excerpt 8b

1

AH supermarket

ANGST&RUIMTE
FEAR&SPACE

VULNERABILITY

HEALTH RISK

Precautions:
- Buy a phone radiation-detection device.
- Send an SMS instead of having a mobile phone conversation.
- Keep conversations short.
- Use a headset.
- Don't give small children mobile phones.
- Don't carry a mobile on your person while recharging the battery.
- Preferably don't make phone calls in the car at all, but when it's unavoidable use a hands-free set in combination with an external antenna.

Active mobile phones are a source of high frequency electromagnetic radiation. The possible health risks will only become apparent with time.

FRAUD

There is growing disparity concerning mobile phones going off, people having loud conversations and careless behavior due to mobile phones.

Each year there is a twenty-five million euro debt due to unpaid mobile phone bills. 80 000 users have an average debt of 350,00 euro.

13.9 million SIM cards

Mobile phone

TEMPORARY TERRITORY   CRIM

TERRORI

Modern technology allows criminals and terrorists to easily falsify travel documents and bank bills, making traditional control systems unreliable.

Mededeling van de politie

Bij alle vervalste bankbiljetten moe
politie inschakelen.
Onze excuses voor het ongemak.

TERRORISM

Your mobile phone is surreptitiously revealing information about finances, shopping habits and lifestyle to outside forces such as enforcement agencies and market researchers.

WESTERN MONEY
UNION TRANSFER

Western Union Money Transfer

The free movement of capital generates global patterns of worker migration.

Footloose capital investment

INTIMIDATION

Night shop

24/7 Economy

SOCIAL FRAGMENTATION

Phone shop

1

VIOLENCE

Your chances of being a victim of abuse or violence on public transportation, is three to four times higher than in the rest of the public domain.

The average Dutch person spends one day a week in traffic or on public transportation.

Public transportation

2

The parking policy is based on the principle that paid parking should serve as a control device.

Parking policy

Private transportation

Mobility and diversification can lead to a loss of social cohesion.

1

POVERTY GAP

The annual turnover of Ford Motor Co. is higher than the GDP of South Africa.

INEQUALITY

LOSS OF SOCIAL COHESION

Parking machine

Parkeerduur onbeperkt, parkeertijd per € 0,50

There are currently 22 000 regulated parking places. By 2010 this must have increased to 50 000.

The City Parking Company makes a profit of 4.8 million euro per year.

INTOLERANCY

NEGATIVE IMAGE
FRAUD
Mr. Smith

TERRORISM   XENOPHOBIA

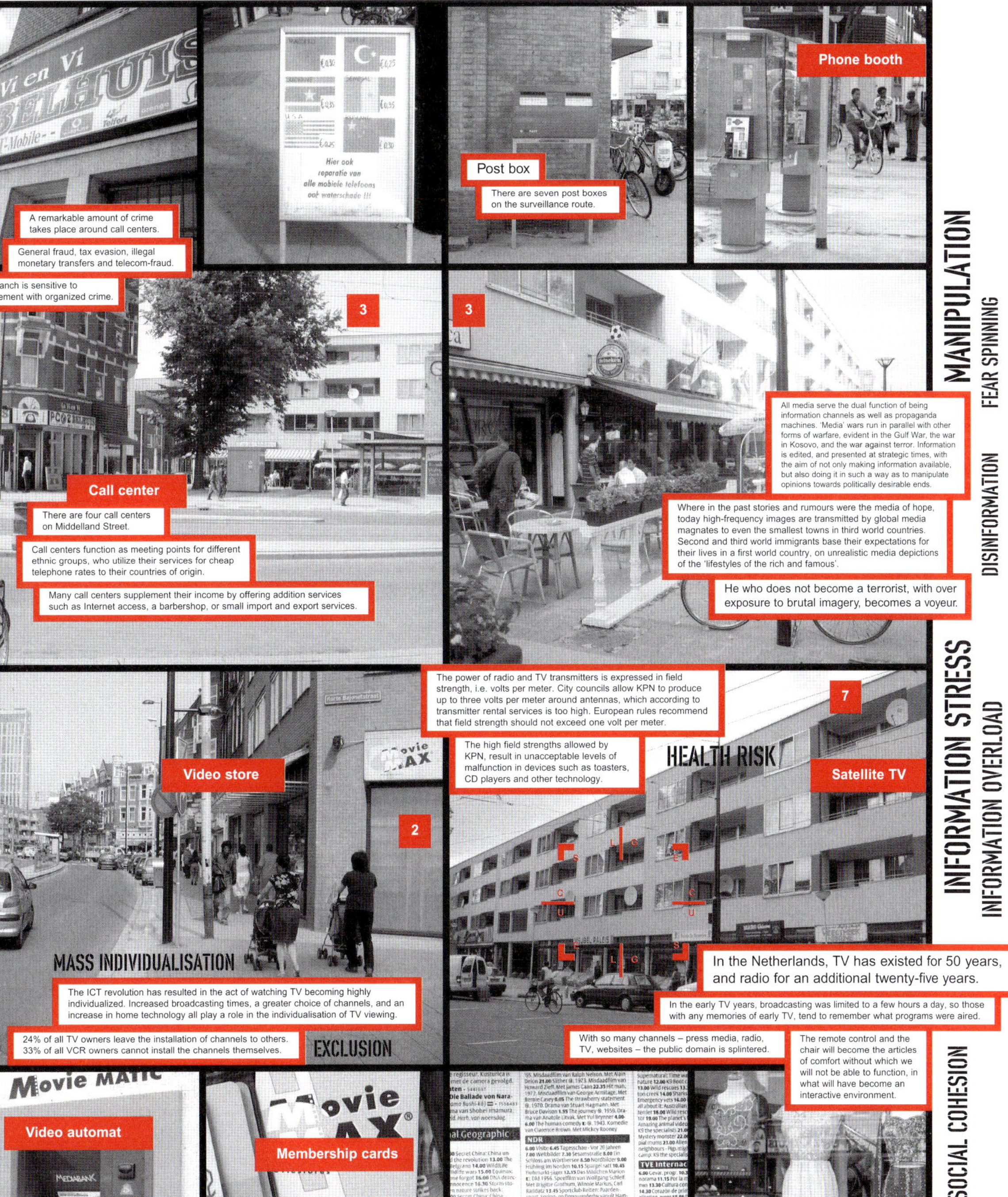

MONOPOLY

VULNERABILITY

GLOBAL MIGRATION
XENOPHOBIA

The NMA suggests that Interpay Nederland B.V. and related shareholders charge unreasonably high PIN rates, and should have competitive agreements in place.

Despite efforts on the part of the International Monetary Fund and World Bank, many countries still suffer from excessive debt, poor financial supervision and management, and weak market shares. The reinforcement of national finances is essential in overcoming this trans-border financial crisis, and calls for global action.

Interpay is the only supplier in the Netherlands of support services for PIN payments. It was established by eight banks and is solely owned by them. Modern technology enables Interpay to process millions of transactions made with bankcards, cheques, credit cards and Chipknip.

Emigration and inequality are connected in many ways. Emigration is partly a reaction to inequality. This inequality can be lessened if emigrants achieve their aims, and accentuated if they don't.

iBoko enables you to send money online and receive money from over 700 000 ATMs in 170 countries.

75% of Middelland Street's inhabitants are foreign.

The first ATM was installed in the Manhattan Chemical Bank in 1969.

ATM

In 2000 there were one million ATMs worldwide. There are currently 8 000 ATMs in the Netherlands alone.

On the surveillance route there are thirteen ATMs

As the demand for this service increases, even more ATMs are being installed.

4

Excerpt 8c

One cash dispenser can effectively do the work of thirty-seven bank tellers.

In 2001 the number of banks in the Netherlands fell by six percent to 5 700.

Unemployment in Rotterdam rose from 3.2% in 2003 to 10.3% in 2004.

The rise of the consumer society and increased material wealth has generated a new kind of uncertainty, as it comes paired with a new ideal: a practical belief in the fundamental equality of all men and in our unlimited capability to achieve whatever we set out to do.

EXPLOITATION

LOSS OF SOCIAL STANDING

UNCERTAINTY

POVERTY GAP

UNEMPLOYMENT

FRAUD CRIME

Angst in de discotheek   Mr. Smith

MONEY LAUNDERING

Skimming

High-tech criminals can copy your PIN card and code, enabling them to gain access to your account without your knowledge. Any transactions they do, will appear legitimate. It is only you, who may notice these discrepancies in your bank statement.

Yearly, an average of one billion dollars is laundered worldwide. An expanding global monetary network facilitates this to a large degree.

The glue trick

Ghost transactions

Pick pocketing   Mugging

Protecting your ATM card:
- Keep your card in a safe place to avoid damage.
- Memorize your Personal Identification Number (PIN). Never write the PIN down on anything in your wallet or on the card itself.
- When selecting a PIN, avoid numbers and letters that relate to your personal information. For example, don't use your initials, birthday, telephone or Social Security number. If you have such a number, contact your bank and get a new PIN issued.
- Immediately report a lost or stolen card to your financial institution.
- To help guard against fraud, keep your ATM receipts until you check them against your monthly statement.

Common sense usage tips:
- Observe your surroundings before using ATM. If the machine is obstructed from v or poorly lit, visit another ATM.
- Take a friend with you - especially at nigh
- Have your card out and ready to use
- Shield the screen and keyboard so anyor waiting to use the ATM cannot see you e your PIN or transaction amount.
- Put your cash, card and receipt away immediately.
- Count your money later, and always keep your receipt.
- If you see anyone or anything suspicious cancel your transaction and leave immediately. If anyone follows you after making a transaction, go to a crowded, w lit area and call the police.
- hen using an enclosed ATM that requir our card to open the door, avoid letting rangers follow you inside.
- hen using a drive-up ATM, make sure a assenger car doors are locked and indows are up.
- o not leave your car unlocked or engine nning when you get out to use an ATM hile many ATMs are available 24 hours ay, some may be open only during loca usiness hours. To be on the safe side, p your withdrawals ahead of time.
- Check with your financial institution to determine the daily limit of funds that car withdrawn.

Preventative measures to disable skimming:
Piano device, Enhanced card drive jitter, Enhanced card reader surround or Biometrics.

Two developments in the realm of information technology will be of importance regarding an individual's identity in our information society: multifunctional smart cards and biometry. Large scale usage of chip cards will make electronic identities possible in the future. This sheds new light on the notion of identity fraud, because smart cards offer the possibility to create false and pseudo-identities that are difficult to trace in the event of a transaction.

Experts suggest that in order to make using PIN's safe in the future, biometric solutions will have to be implemented. Physical characteristics unique to each person, such as fingerprints, retina and voice structure will be used to safeguard systems. For example, a fingerprint can be read by an ATM and translated into numeric code thereby verifying the identity of the user.

PIN

In the future, the large scale use of chip cards and electronic identities, will result in a wide range of existing, and yet undefined forms of misuse taking place.

In the 70's it was easy to establish someone's identity. Bank tellers were likely to know and recognize their customers. People were more likely to use cash to pay at supermarkets etc. With the introduction of computers into our society, this has changed significantly. Computer transactions now occur more frequently than cash transactions.

ANONYMITY

LOSS OF CHOICE

An increasing number of banks are closing down, due to mergers and the automation of the banking process. Because of this, people no longer have a choice but to withdraw their money from an ATM.

In areas where the population density reaches 5 000 people within a three kilometer radius, a number of basic services are required to be present. Because of the reduction in the number of banks, post offices and municipalities, many citizens in these areas now do not have access to their money.

Clients are no longer able to make withdrawals of less than five hundred euro inside the bank at the bank teller.

DEPENDENCY

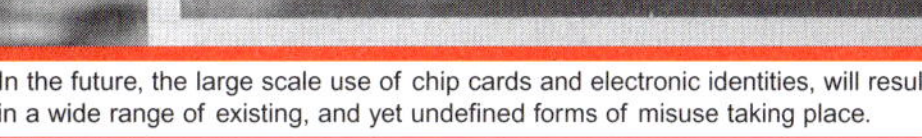

Banks are prohibited from selling customer data. This information is however made available to the Credit Registration Bureau, investigation institutions, pension funds and insurance companies.

Visa keeps expenditure records for two years enabling the generation of customer profiles. This in turn makes it possible to send personalized advertisements with the monthly bills.

In the USA, Visa has plans to sell client information directly to market researchers.

More than the internet surfer, the user of chip, PIN, credit and client-cards, leaves spores behind him. The PIN code leaves the most elaborate spores.

TERRITORIAL BEHAVIOR

LOSS OF PRIVACY

TRADE IN PERSONAL INFORMATION

ne third of the elderly avoid using ATMs.

Elderly people often have trouble remembering their PIN code and writing it down increases their risk of being robbed. To reduce their levels of anxiety and fear of using isolated ATMs, many make use of cash-back facilities at supermarkets instead.

EXCLUSION

In 2050 there will be an additional one million people over the age of 75 compared to the 1.2 million of today.

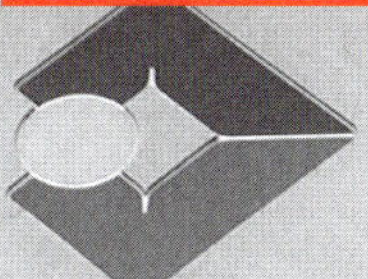

Humanitas Old Age Home

Few ATMs if any are accessible to people in wheel chairs or with limited hand mobility.

German health insurers plan to give citizens a chip-card, which also provides access to fitness centres and swimming pools. The more people keep fit the lower the premium to be paid. Buying cigarettes, on the other hand would lead to an increase in their monthly premium.

Fear at the Discotheque   Mr. Smith

TERRITORIAL BEHAVIOR

SIGNS OF DECAY

STREET GANGS

Excerpt 8d

DATA-MINING

SIGNS OF DECAY

VULNERABILITY

CRIME

Mark and register all valuables kept in your car, such as the make of car-radio or mobile phone, including type and serial number. In this way the chance of retrieving your goods after a theft are increased. You can record this information on the free registration forms provided by the police. Always include your post code and address.

For a few hundred euro, most brands of car keys can be bought on the internet allowing thieves to break into any car within seconds.

Car thefts are not only carried out by local professional criminals, but also by East European crime syndicates.

auto is leeg!
Empty

On average one car is stolen in the Netherlands every two minutes.

Car theft

The socio-economic effects related to drug trade, such as shoplifting, burglary and prostitution cost national governments a vast amount of money

6

The police have limited means to tackle drug trading. Preventative measures such as extra policing and surveillance cameras only result in a shift of the drug-scene from one part of a city to another

Coffee shop

Conservative political speakers swear allegiance to an imaginary regime, whereby so called morality & propriety, discipline & order once reigned. They see the decline of the world over the last two hundred years as a result of emancipatory waves and the fall of formal authorities.

5

CRIME

Street gangs

Gang wars can no longer be attributed to traditional motives such as the historical class struggles between black and white. In these small scale 'civil wars', there are no winners.

People either recognize their sameness to others or their differences. Focusing on the differences creates intolerance, to the point where even the smallest details become instigators of hatred. For example, what football club you support, the clothes you wear, or the language you speak.

The Netherlands is the world leader in the production and distribution of XTC.

TERRORISM

Global policies are needed to curb the international production of drugs. If not integrally tackled, drug producers will just shift from country to country, because drugs are highly profitable and easily transportable.

With a turnover of approximately 150 billion dollars and about 200 million users, the illegal drug market is the biggest clandestine market in the world.

Narcotic profits often feed the accounts of terrorists or rogue-states.

There are five phone booths on Middelland Street.

There are a total of 18 500 phone booths in the Netherlands.

KPN is obliged to provide one phone booth for every 5 000 citizens.

Phone booth

5

8

kpn telecom

Datbank KPN: time and duration of call, account number, account overview, identity of caller, origin and destination of call are registered and saved for six months.

Often the booths are dirty or vandalized. KPN's profits do not cover the damage.

Due to the rapid growth in mobile phone use, KPN is reducing the number of Dutch phone booths.

The reduction of 3 400 phone booths represents the loss of 140 KPN jobs.

Toko

TERRITORY

The globalisation of food production systems is destroying the diversity of local food cultures and produce economies.

UNEMPLOYMENT

Drug runner

It is possible to trace a phone call to a phone booth, but the identity of the caller remains anonymous.

Intercom door

Ethnic minorities make up ten percent of the Dutch population, leading to mixes of lifestyles, eating habits and consumption patterns. This is evident in the increasing number of shops with indigenous products.

Based on current trends, the prognosis is that in the coming twenty years, more Chinese men will search for wives in foreign countries and simultaneously study there.

The ministry has agreed to intensify policies against drug couriers. The existing policies lead to conflicts in the legislative process.

Stricter legislation limiting the sale of soft drugs from coffee shops has resulted in more drug trade on the streets.

CRIME

BRAIN DRAIN   XENOPHOBIA   339 Immigrants enter the Netherlands per day.

I want to thank the following people/institutes who inspired and influenced me. Sometimes I used fragments of their texts, quotes or statements freely to make the reconstruction of my fearspaces. If I have forgotten someone, please realize that this is neither deliberate nor negligent, but a side effect of my psychosis. AMO; Antenna; Archis; de Architect; Automation guide; B4U Freeband; Paul Bairoch; Benjamin Barber; Richard Barbrook; John Perry Barlow; Jeremy Bentham; Joost Blokzijl; Blueprint; William Bogard; Alain de Botton; Arianna Bove; Bradford; Philip Brey; Lester Brown; Bureau for Credit Registration; George W. Bush; CNN; Sophie Calle; Manuel Castells; Lieven de Cauter; Center for Research and Statistic; Central Bureau for Statistics; Paul Chevigny; Im Sik Cho; Christopher Dandeker; Council for Protecting Citizen Data; Cybergeography research; Mike Davis; Peter Dicken; Martin Dodge; Natanya Doepel; Eindhovens Dagblad; Anton Ekker; Henk Elffers; Elsevier; Erik Empson; Okwui Enwezor; Hans Magnus Enzensberger; Factomedia; Roger Fawcett-Tang; Andrew Feenberg; Edwin Feldmann; Fietsersbond; Het Financieele Dagblad; Sander Flight; John Friedmann; GEP2000 UNEP; Globalinfo; Stephen Graham; Tijs Goldschmidt; De Groene Amsterdammer; T.J. Golder; HP/De Tijd; Stuart Hall; Michael Hardt; Yvonne van Heerwaarden; Lynn Hershman Leeson; Ankie Hoogvelt; Ikobo; Infodrome; Celine Jeanne; KPN; Joshua Karant; Douglas Kellner; John F. Kennedy; Max Kisman; Naomi Klein; Bert v.d. Knaap; Kondratiev; Minister van Justitie; J. Kruize; Thomas V. Levin; Geert Lovink; David Lyon; Simon Marvin; MasterCard; Mediafact; Thomas Misa; Chantal Mouffe; Municipality Amsterdam; Municipality Rotterdam; NMa; NOS; NOVIB; NRC Handelsblad; Netherlands Institute for the Study of Crime and Law Enforcement; Antonio Negri; Next Architects [Excerpt 15c-d]; Michiel van Nieuwstadt; Erica Nijburg; Nu website; George Orwell; OOV; Research School Safety and Security in Society; William Owen; Pvda; Het Parool; Pico website; Planet; Rotterdam Police website; Mark Poster; Thomas Pynchon; Ravage; Research Institute Leiden; W.J. de Ridder; Mirjam de Rijk; Ned Rossiter; Marijn Schenk; Suitbert Schmit; SEC; SMO; Security Management; Vandana Shiva; Uno Smith; Bert Steinmetz; Stepnet; The Social and Cultural Planning Office of the Netherlands; Bob Sutcliffe; Swann; TICV; TNO; TU Delft; Telecompas; Telecommagazine; De Telegraaf; Telematica Institute; Siebe Thissen; Harm Tilman; Three Angels Broadcasting; Trouw; Tumult; USD congress concerning XTC; United Nations General Assembly; University of Amsterdam; University of Groningen; Vahan H. Haji-Petros & ECRG; Van Traa-team; Paul Virilio; de Volkskrant; McKenzie Wark; Wikipedia; Wired; James Wolfensohn; Goetz Wolff; Yearbook IT and Society 2003; ZDNet.

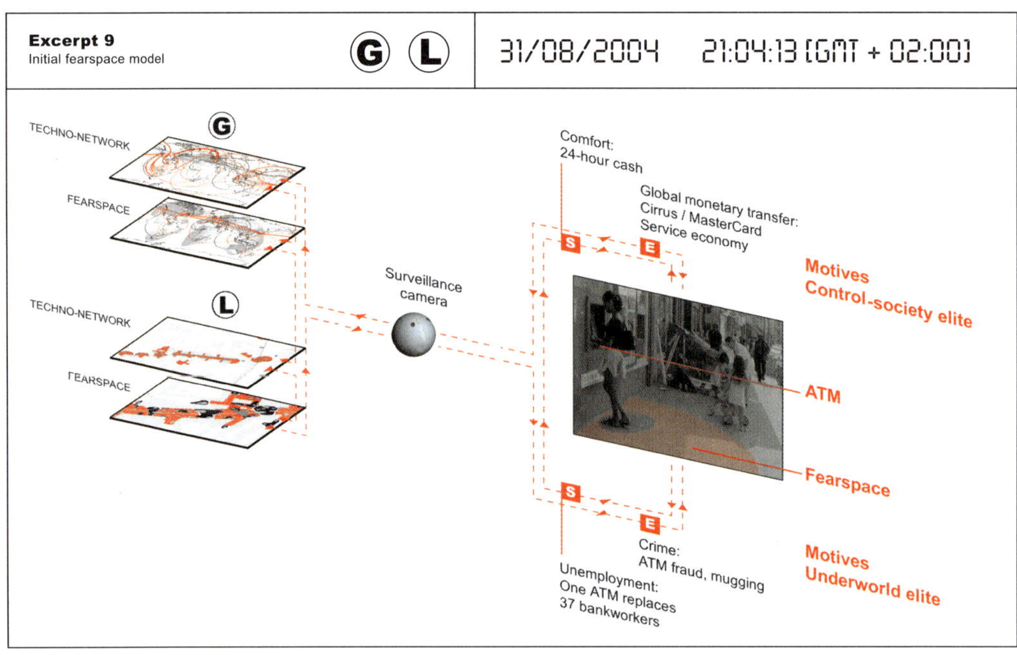

"SURVEILLANCE PRESENTS RISKS OF WHAT ULRICH BECK CALLS THE 'RISK SOCIETY' IN WHICH THE MODERN INDUSTRIAL PRODUCTION OF 'GOODS CARRIES WITH IT A LESS OBVIOUS PRODUCTION OF 'BADS' IN UNFORESEEN SIDE EFFECTS AND IN ENVIRONMENTAL DESPOLIATION. THE RISK SOCIETY ARISES IN 'AUTONOMISED MODERNIZATION PROCESSES WHICH ARE BLIND AND DEAF TO THEIR OWN EFFECTS AND THREATS, RESONATING IN THE CURRENT SURVEILLANCE SOCIETY." (DAVID LYON - MODERNITY AND TECHNOLOGY)

In each case, we see how the previously defined 'techno-filter' has been applied by Smith in making these diagrams. For example, in [excerpt 9], a photo is shown, in which a normal person only sees a typical street setting – but where Smith's psychosis would fabricate, (1) the targeting of a technological interface (in this case an ATM[68]); (2) the socio-economic evaluation of the interface under both control-society and underworld motives; (3) the connection of these motives to a 'control technology' (in this case a surveillance camera); (4) the translation of this knowledge into an integral global and local technological network; (5) the conversion of this into global and

In alle gevallen zien we hoe Mr. Smith zijn vooraf bepaalde 'technofilter' gebruikt om deze schema's te maken. In [excerpt 9] wordt bijvoorbeeld een foto getoond waarin een normaal persoon een doodgewoon straattafereel zou zien, maar die door de psychotische Mr. Smith wordt geïnterpreteerd via (1) het focussen op een technologische interface (in dit geval een betaalautomaat[68]); (2) de sociaal-economische evaluatie van deze interface vanuit de motieven van zowel de onderwereld als de controlemaatschappij; (3) het verband tussen deze motieven en een 'controletechnologie' (in dit geval een bewakingscamera); (4) de vertaling van deze kennis naar een geïntegreerd mondiaal en lokaal technologisch netwerk; (5) de omzetting van dit netwerk in mondiale en lokale angstruimtes en

---

Excerpt 10
ATIC definitions

Fearspace	Fearspace
A AGONISTIC — Agonistic space represents a domain where the behaviour of an individual is (un)consciously, subject to any act or system of implicit control, such as surveillance cameras, satellite systems, spyware, membership cards and global monetary transfers.	**I** INTERACTIVE — Interactive space represents a domain where the behaviour of an individual is influenced by the bottom-up social control of an individual, or groups of individuals towards any other individual or groups of individuals such as chatboxes, internetforums, webcams and group conduct in public space worldwide.
T TERRITORIAL — Territorial space is an area of self-control, caused by conditions under which an individual is urged to initiate or create his own [environmental bubble] to defend himself from control systems. This capsularity can be both physical, such as shielding your pincode at an ATM, or virtual, with a fire-wall. At the global scale these are spaces such as money laundering paradises and free zones.	**C** CONFLICT — Conflict space is a domain in which the presence of an individual and his behaviour is subject to the approval of acts and systems of direct and explicit control such as guards, security cameras (CCTVs) and global coalitions such as Opec, G8, the Arab League and Organistation of Islamic Conference.

Excerpt 11
Neurological model of fear

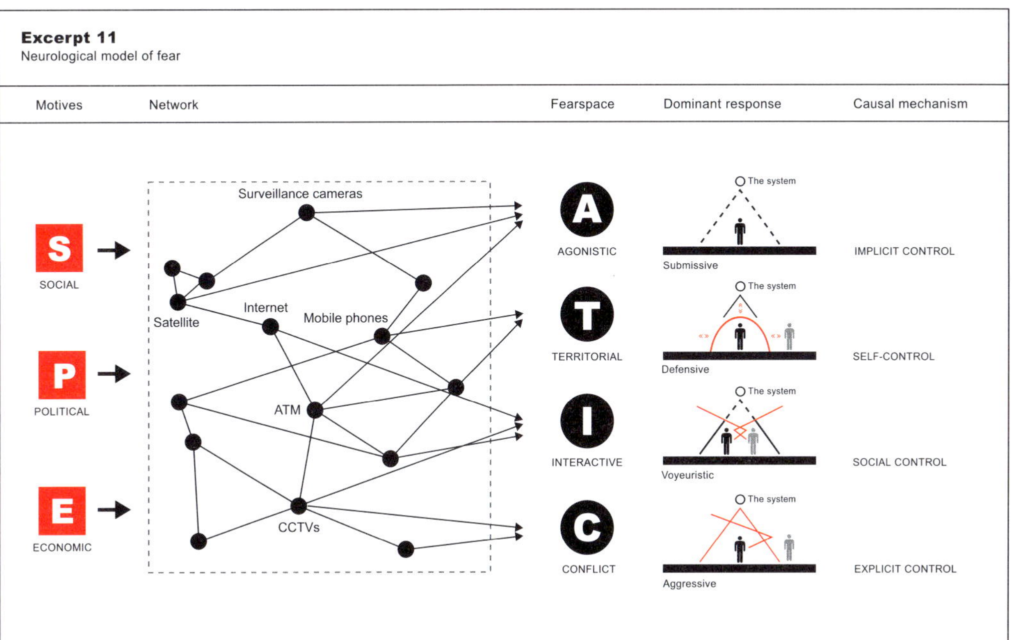

local fearspaces; and (6) the ethical judgement of the overall system. This produces Smith's anxiety. In [excerpt 10] we see a more generalized image of Smith's psychosis, where socio-economic motives are transferred and controlled through technological networks, by which four types of fearspace are effectuated; Agonistic, Territorial, Interactive and Conflict (ATIC). In his obsession to discover lateral relations between things [excerpt 11], Smith observes that a direct link exists between technology (cause) found in public space, and his fourfold (ATIC) spatial anxiety (effect). He believes this relationship is a product of the existing system.

(6) de ethische beoordeling van het totale systeem. Hierdoor ontstaat Smith's angststoornis. In [excerpt 10] krijgen we een algemener beeld van Smith's psychose, waarbij sociaal-economische motieven worden overgebracht en beheerst via technologische netwerken, waardoor vier verschillende typen angstruimtes tot stand worden gebracht: de Strijdlustige, Territoriale, Interactieve en Conflictueuze angstruimte (STIC). Door zijn dwangmatige neiging dwarsverbanden tussen dingen bloot te leggen [excerpt 11], neemt Mr. Smith een rechtstreeks verband waar tussen de technologie (oorzaak) die in de publieke ruimte wordt aangetroffen en zijn vierledige (STIC) ruimtelijke angst (gevolg). Hij gelooft dat dit verband een product is van het bestaande systeem.

Excerpt 12
Social development cycle

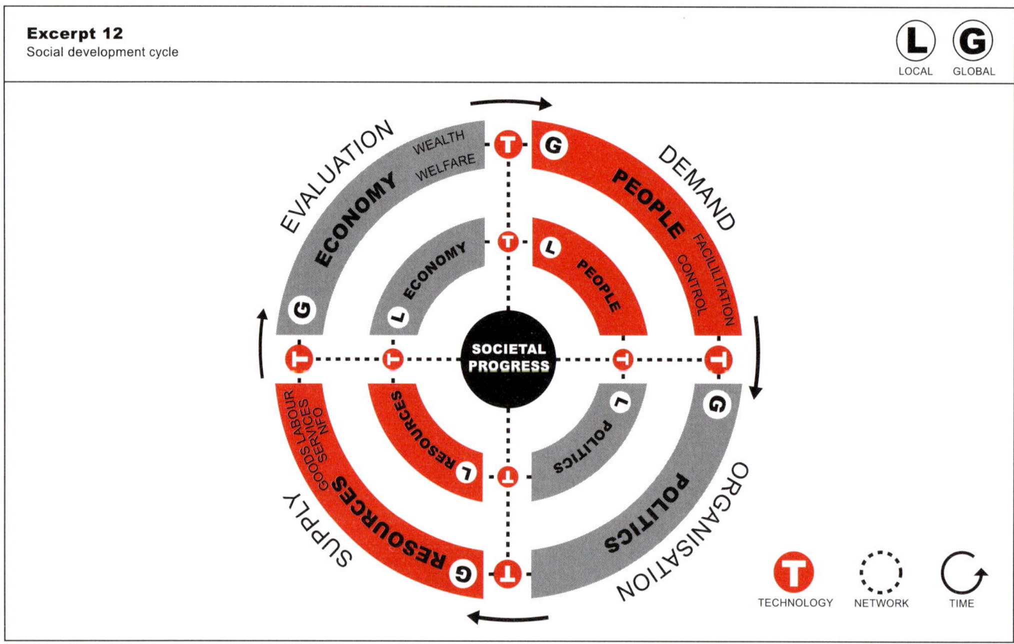

"THE BIGGEST DANGER IS TO PRETEND NOT TO SEE WHAT IS HAPPENING; ONLY THE INVENTION AND PRACTICE OF A NEW DEMOCRACY OF THE WORLD CITIZENS WILL PROTECT US FROM DEVASTATION. WE NEED A DEMOCRACY WHICH CAN STAND UP TO THE MONSTER WHICH PRESENTS ITSELF." (NEGRI, 2002)

C. PATIENT'S CONCEPTUAL MODEL OF FEARSPACES

In the last part of our diagnosis, we introduce Smith's elaborate model of fearspaces. Firstly, it is vital to see how he conceives contemporary human development. In [excerpt 12] Smith illustrates the progression of society as a socio-economic process made up of four segments (demand, organization, supply and evaluation). This system is divided into local and global rings and all items are interconnected by technological networks. From this Smith demonstrates today's society as an intricate socio-economic model

C. HET CONCEPTUELE MODEL VAN ANGSTRUIMTES

In het laatste deel van onze diagnose introduceren we Smith's uitgewerkte model van de angstruimte. Essentieel is in ieder geval zijn kijk op de hedendaagse ontwikkeling van de mensheid. In [excerpt 12] illustreert Mr. Smith de vooruitgang van de maatschappij als een sociaal-economisch proces dat uit vier segmenten bestaat (vraag, organisatie, aanbod en evaluatie). Dit systeem kan worden onderverdeeld in lokale en mondiale ringen, waarbij alle items onderling verbonden zijn via technologische netwerken. Aan de hand hiervan laat Mr. Smith de hedendaagse maatschappij zien als een complex sociaal-economisch model

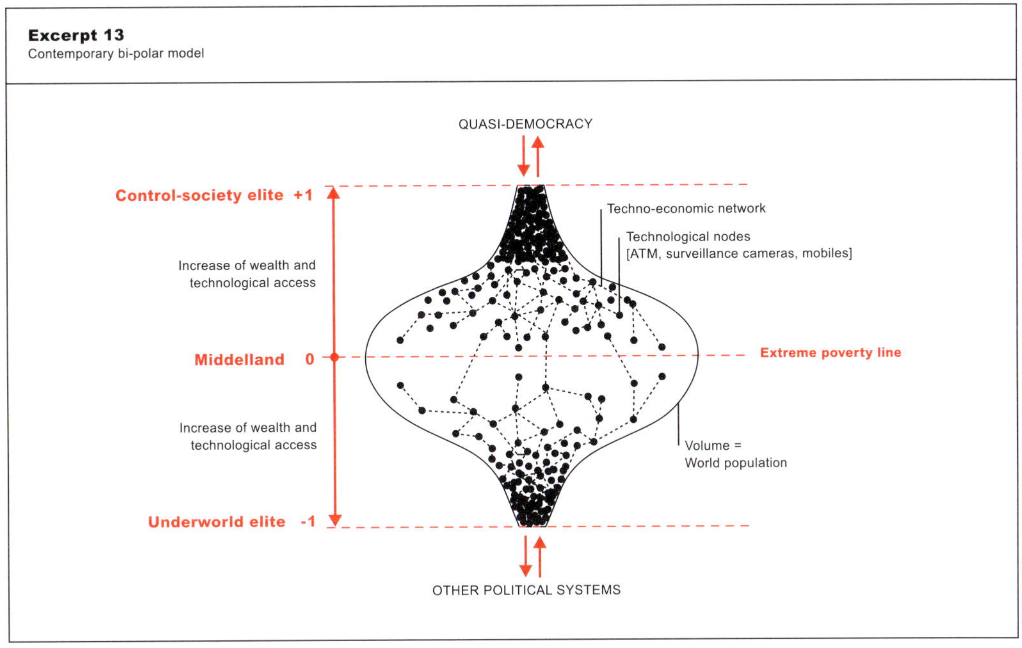

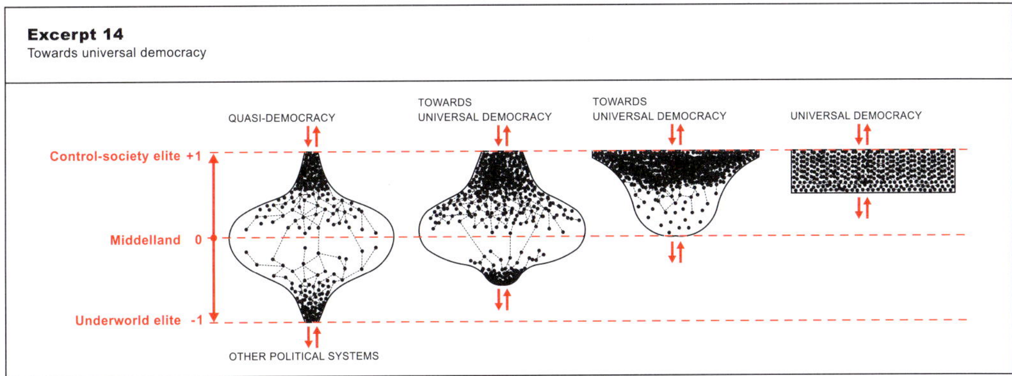

[excerpt 13] that has been shaped by the current political climate. Two poles of wealthy minority are illustrated, namely the control-society (+) and underworld (-). Between these poles, the poor majority is found, forming the bulk of the shape. The midline expresses extreme poverty (0) and the closer people are to the poles, the higher the financial and technological access. This is represented by the increase of black dots (nodes) and lines (linkages). According to Smith, the tension caused between an incomplete democracy and reactionary political or anarchic systems, gives rise to this unjustifiable model.

[excerpt 13] dat is gevormd door het huidige politieke klimaat. De tegenpolen van de twee rijke minderheden worden getoond, die van de controlemaatschappij (+) en de onderwereld (-). Tussen deze polen vormt de arme meerderheid de bulk. De middellijn staat voor extreme armoede (0). Hoe dichter iemand zich bij een van beide polen bevindt, des te meer toegang heeft hij tot financiële en technologische middelen. Dit wordt weergegeven met een toenemend aantal zwarte stippen (knooppunten) en lijnen (verbindigen). Volgens Mr. Smith leidt de bestaande spanning tussen onvolledige democratieën en politieke of reactionaire systemen tot dit niet te rechtvaardigen model.

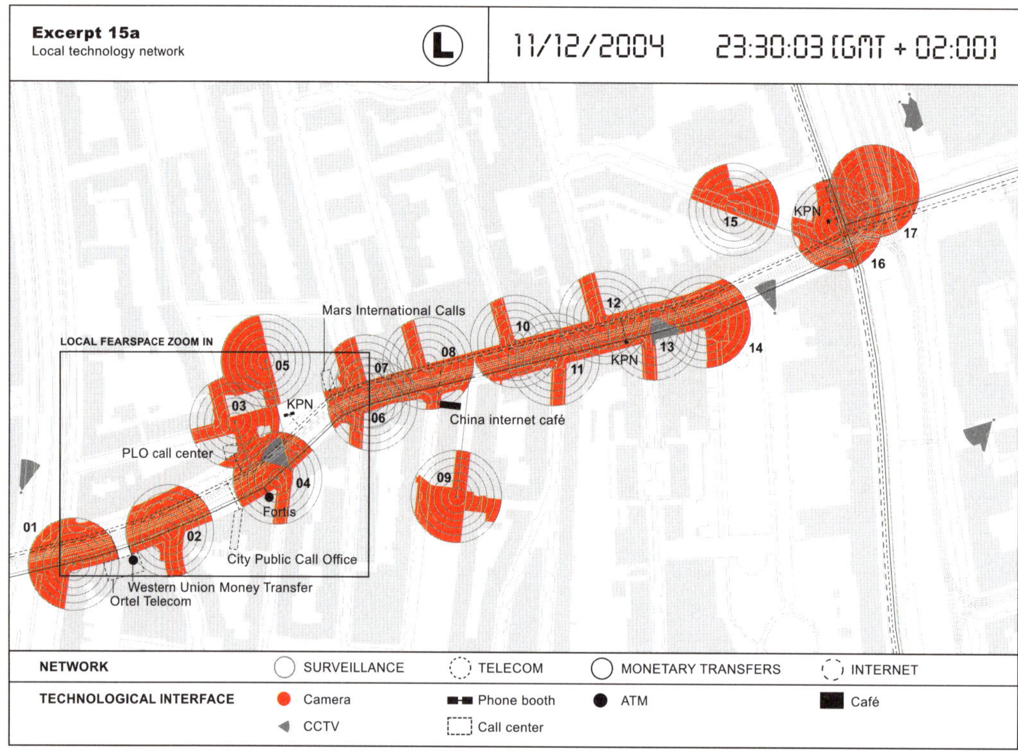
Excerpt 15a
Local technology network
L
11/12/2004   23:30:03 (GMT + 02:00)
15
KPN
17
16
Mars International Calls
12
10
08
07
06
05
11
KPN
13
14
LOCAL FEARSPACE ZOOM IN
03
KPN
09
PLO call center
China internet café
04
Fortis
01
02
City Public Call Office
Western Union Money Transfer
Ortel Telecom
NETWORK
SURVEILLANCE
TELECOM
MONETARY TRANSFERS
INTERNET
TECHNOLOGICAL INTERFACE
Camera
Phone booth
ATM
Café
CCTV
Call center

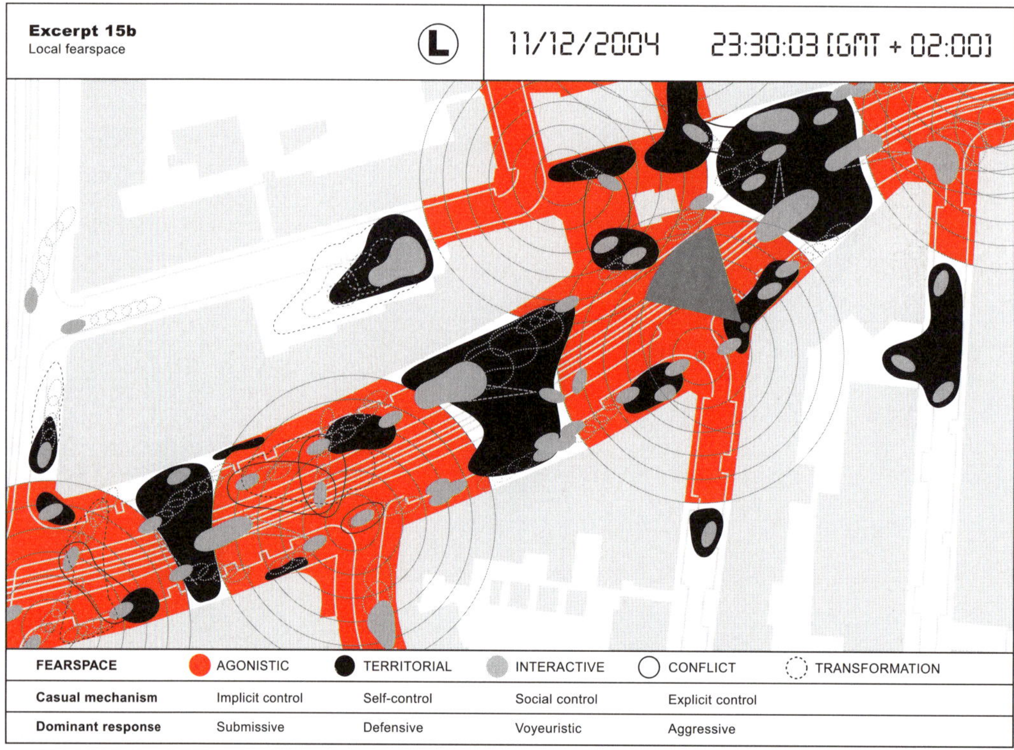
Excerpt 15b
Local fearspace
L
11/12/2004   23:30:03 (GMT + 02:00)
FEARSPACE
AGONISTIC
TERRITORIAL
INTERACTIVE
CONFLICT
TRANSFORMATION
Casual mechanism
Implicit control
Self-control
Social control
Explicit control
Dominant response
Submissive
Defensive
Voyeuristic
Aggressive

Excerpt 15c
Global technology network

(G)   11/12/2004   23:30:03 (GMT + 02:00)

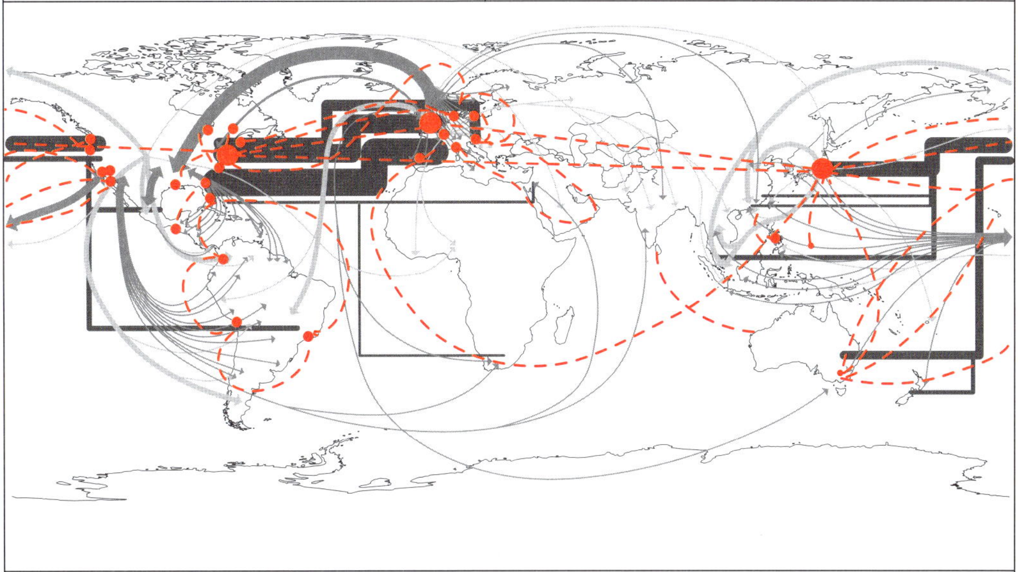

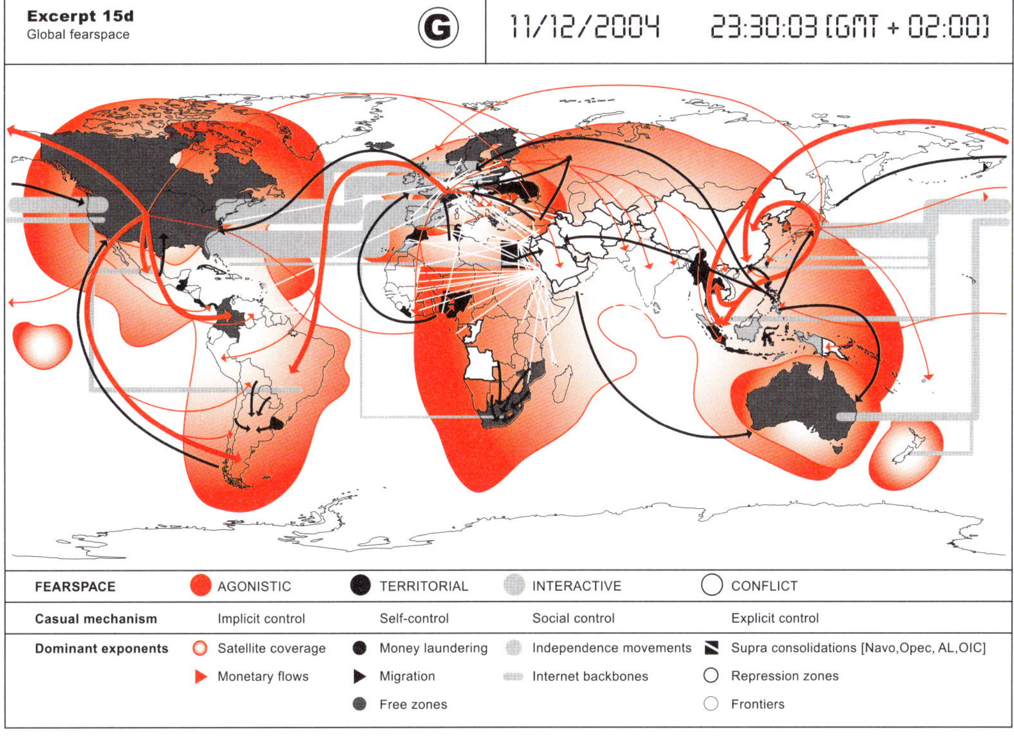

Excerpt 16a
Elaborated spatio-temporal model
G L
11/12/2004   23:30:03 (GMT + 02:00)
SPATIAL
POLITICAL
TECHNO-NETWORK
FEARSPACE
TECHNO-NETWORK
FEARSPACE
DEMOCRACY ?
OTHER SYSTEMS ?
SOCIO-ECONOMIC
Input / Output
Techno-economic network
+ 1 Control-society elite
0 Middelland
Volume = World population
- 1 Underworld elite
TIME

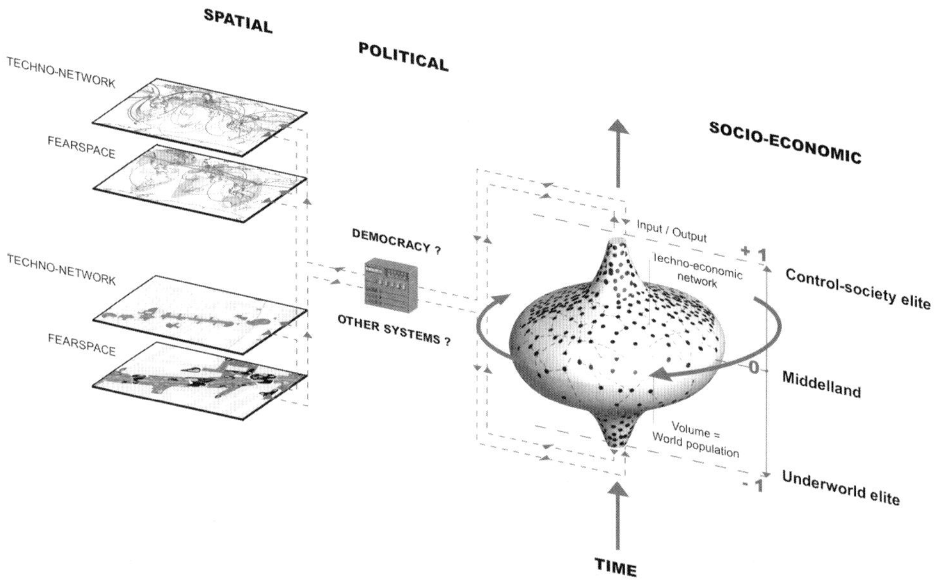

Excerpt 16b
Elaborated spatio-temporal model
G L
11/12/2004   23:30:03 (GMT + 02:00)
SPATIAL
POLITICAL
TECHNO-NETWORK
FEARSPACE
TECHNO-NETWORK
FEARSPACE
DEMOCRACY ?
OTHER SYSTEMS ?
SOCIO-ECONOMIC
Input / Output
Techno-economic network
+ 1 Control-society elite
0 Middelland
Volume = World population
- 1 Underworld elite
TIME

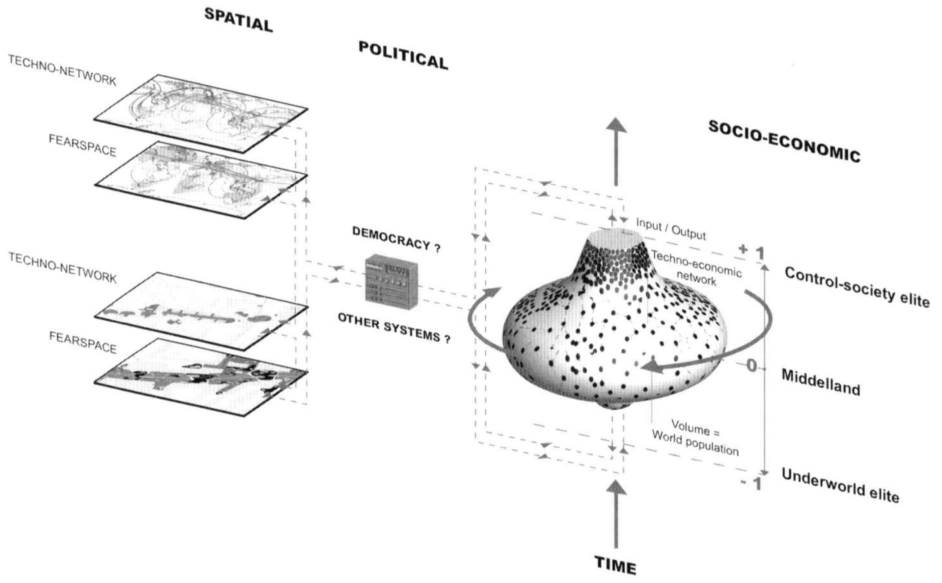

Excerpt 16c
Elaborated spatio-temporal model   (G) (L)   11/12/2004   23:30:03 [GMT + 02:00]

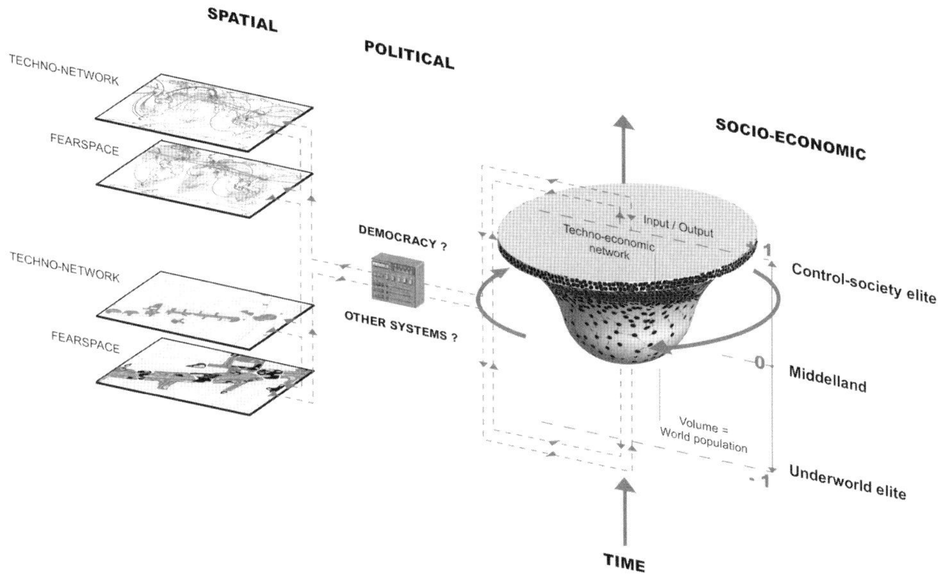

Excerpt 16d
Elaborated spatio-temporal model   (G) (L)   11/12/2004   23:30:03 [GMT + 02:00]

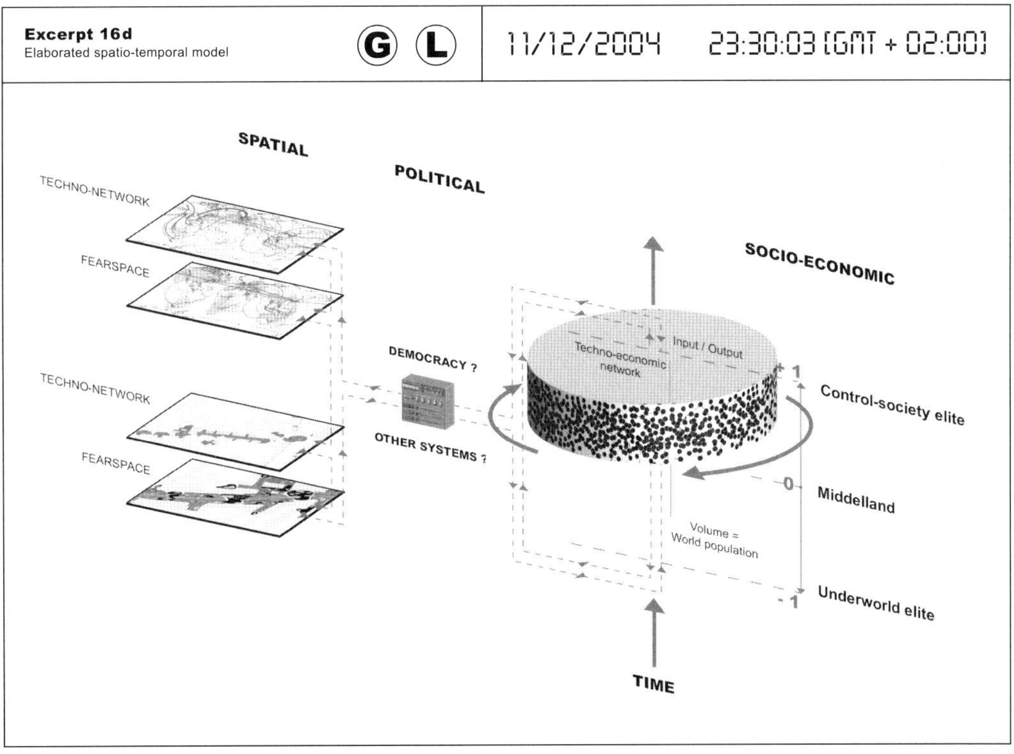

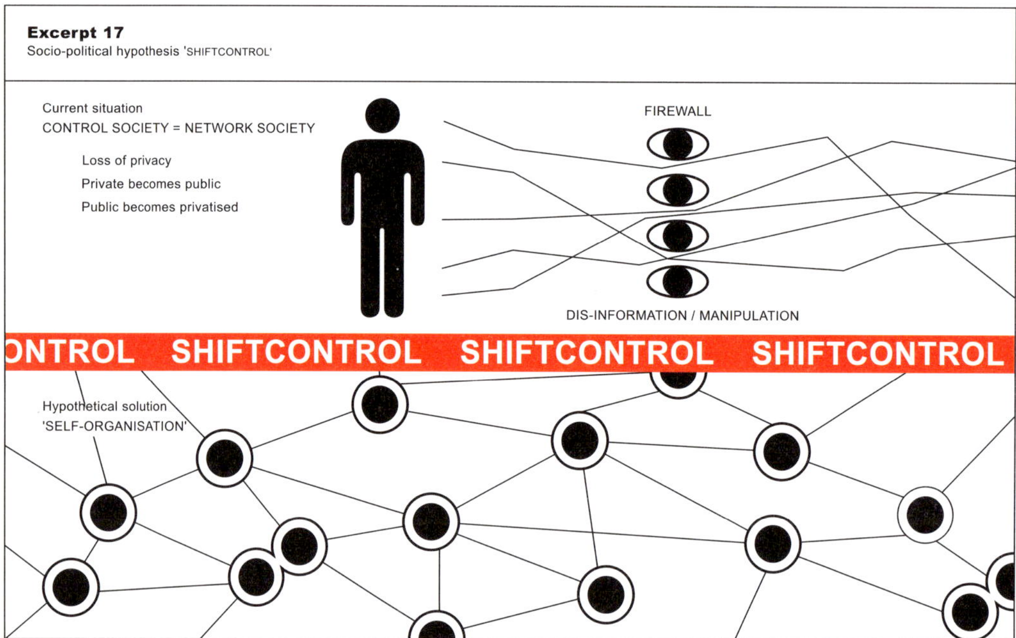

“THE NETWORKS ARE INVENTED, OWNED AND REGULATED BY A MINORITY OF COMMERCIAL, IDEOLOGICAL AND POLITICAL INTERESTS. NETWORKS INFILTRATE OUR SPACES AND BECOME THE ULTIMATE FOLD BETWEEN PRIVATE AND PUBLIC WORLDS. THIS INTERFACE FUNCTIONS AS A FIREWALL, MONITORING AND REGULATING THE PERMISSIBLE FLOWS, OPERATED BY VARIOUS GATEKEEPERS. THE ‘NETWORK SOCIETY’ IS THE ‘CONTROL SOCIETY’. IT IS THERE-FORE URGENT THAT WORLDWIDE SOCIAL AND POLITICAL NETWORKS REDEFINE POLICIES CONCERNING ‘CONTROL’ AND ‘SELF-ORGANIZATION’ HEREBY ENABLING ACCESS AND EMANCIPATING THE MAJORITY.” (SMITH, 2004)

In [excerpt 14] (p. 39), Smith proposes transitory steps towards a universal democracy – although aware of the improbability. By combining [excerpt 9,13 and 15a-d] the most elaborate of Smith's fearspace studies is generated [excerpt 16a-d]. We see how the previously described schemes lead to the production of integral, local and global fearspaces. The concept of time has been introduced, whereby Smith implies the transformation of the model along the time-axis, leading to the model's transition.

We are pleased that you are willing to engage with this severe case and we hope your expertise in TDP will lead to a proper prognosis and possible treatment for Mr. Smith. Below, we provide you with some troubling questions that Smith has posed to us, which we are unable to answer. These might assist your assessment.

In [excerpt 14] (p. 39) suggereert Mr. Smith stappen om te komen tot een universele democratie – al twijfelt hij aan de haalbaarheid. Door [excerpt 9, 13 en 15a-d] te combineren wordt het meest omvattende van Smits angstruimteonderzoeken gegenereerd [excerpt 16a-d]. We zien hoe de eerder beschreven processen leiden tot het ontstaan van integrale, lokale en mondiale angstruimtes. Het concept tijd wordt geïntroduceerd; aan de hand van een tijdsas suggereert Mr. Smith een verandering van het model in de loop van de tijd.

Het verheugt ons dat u beiden bereid bent zich met dit ernstige geval bezig te houden. We hopen dat u met uw gespecialiseerde kennis en ervaring met TDF tot een duidelijke prognose zult komen en mogelijke behandelmethoden voor Mr. Smith kunt voorstellen. Hieronder volgen enkele brandende vragen die Mr. Smith ons heeft voorgelegd, maar waarop wij geen antwoord konden geven. Deze zijn wellicht van belang voor uw onderzoek.

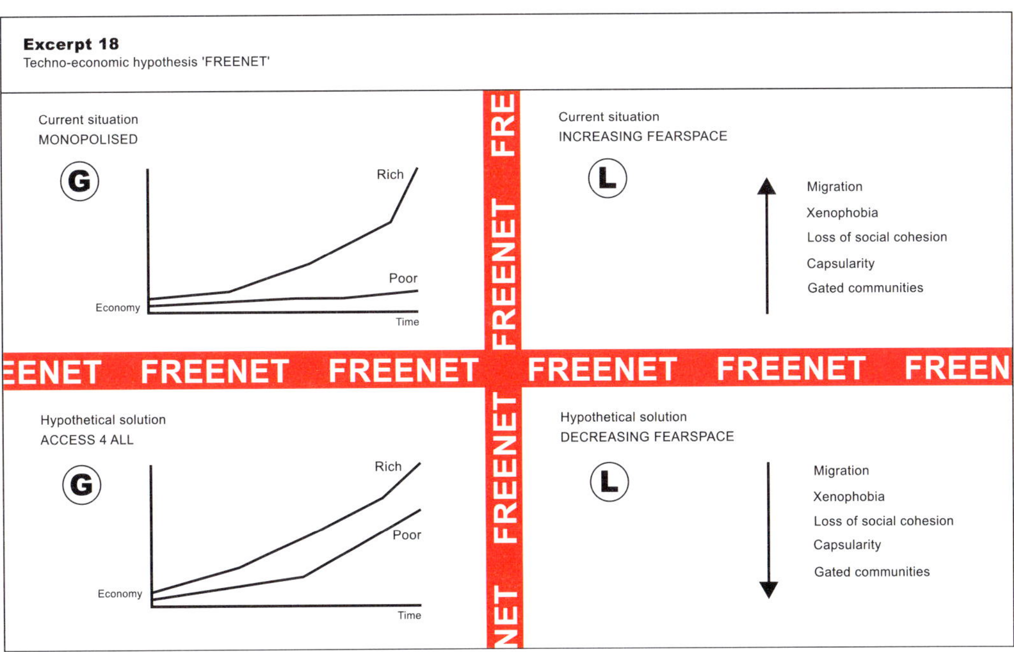

"MONOPOLIZED NETWORKS INCREASINGLY INFILTRATE AND DETERMINE THE SHAPE OF OUR LIVES AND OUR FEARSPACE. GATED COMMUNITIES, CAPSULE-FORMATION AND HIGH SECURITY SPACES ARE THE RESULTANT OF THIS. IT IS THEREFORE URGENT THAT NEW NETWORK TECHNOLOGIES SUCH AS 'NEXT GENERATION NETWORKS', 'OPEN SOURCE' AND 'FREE SOFTWARE MODELS' ARE DEVELOPED BY THE MAJORITY. THIS MUST PROVIDE ENABLING AND INCREASED 'ACCESS 4 ALL' THEREBY LIBERATING PROGRESS OF SOCIETIES AT LARGE." (SMITH, 2004)

1. Are my local fearspaces highly related to global forces?
2. If the system is not altered, is it possible that my fearspaces will maximize?[69]
3. I imagine my fearspace can be transformed by the invention of a genuine, worldwide democracy, (shiftcontrol [excerpt 17]) and the development of enabling technologies[70] (freenet[71] [excerpt 18]). Is this plausible or are there more factors to be considered?
4. Is my fear spatial, ideological or both?
5. According to the philosopher Thomas Hobbes, fear cannot be eradicated and instead serves as a vital generator of societal transformation. Therefore, will my fearspaces only be transformed into new ones?

1. Houden mijn angstruimtes verband met mondiale krachten?
2. Als het systeem niet verandert, kunnen mijn angstruimtes dan tot maximale grootte uitgroeien?[69]
3. Ik stel mij zo voor dat mijn angstruimte kan worden veranderd door de uitvinding van een echte wereldwijde democratie (shift-control [excerpt 17]) en de ontwikkeling van faciliterende technologieën[70] (freenet[71] [excerpt 18]). Is dit plausibel of zijn er nog andere factoren om rekening mee te houden?
4. Is mijn angst ruimtelijk, ideologisch of beide?
5. Volgens de filosoof Thomas Hobbes kan angst niet worden uitgebannen en vervult ze een essentiële rol als aanjager van maatschappelijke verandering. Betekent dit dat mijn angstruimtes slechts zullen worden omgevormd tot nieuwe?

PROGNOSIS
DR. BARBER AND DR. KARANT

Thomas Hobbes was no stranger to fear. In 1651, writing amidst an epoch of intense civil strife, he described human existence as an atomistic 'war of all against all'. Drawing upon the political instability of his age, Hobbes contended that man was by nature hostile and competitive, a creature that struggled to satiate individual desires no matter the means. Given this putatively violent nature, he concluded that only one political remedy could protect citizens from themselves and each other: monarchic rule was his strong-fisted solution to the imminent threat of physical harm so rampant amongst creatures naturally prone to mutual destruction.

More than a century after the publication of Hobbes' Leviathan, Jean-Jacques Rousseau argued precisely to the contrary. Far from an unruly conflict, Rousseau understood man's natural state as one of simple self-love and pitié. Life was characterized not by fierce independence but by interdependence – by the shared universal instinct for self-preservation, and the intuitive sense that an individual's livelihood was deeply connected to that of his peers.

Deeply mistrustful of human nature, Hobbes embraced authoritarianism as the only possible means of saving mankind from itself. Yet as Rousseau insisted, Hobbes' prescription was fundamentally wrong; he ascribed artificial qualities (such as vain self-interest and violent competitiveness) to naturally benevolent creatures. Because, Rousseau countered, man was perverted by his environment – by illegitimate social and political relations which had corrupted an otherwise innocent species – redeeming citizens demanded reforming their institutions. Where Hobbes turned to the sword for protection and stripped subjects of their political will, Rousseau championed an active citizenry cultivated through a system of popular sovereignty drawn according to democratic principles of freedom, liberty and equality under law. Against Hobbes, Rousseau argued that the inequalities and oppressive relations so destructive of human welfare might be redressed only through greater involvement.

The state of our contemporary world may

PROGNOSE
DR. BARBER EN DR. KARANT

Voor Thomas Hobbes was angst een vertrouwd begrip. In 1651, in een tijdperk van hevige burgerlijke conflicten, beschreef hij het menselijk bestaan als een atomistische 'oorlog van allen tegen allen'. Gesteund in zijn opvattingen door de heersende politieke instabiliteit beweerde Hobbes dat de mens een van nature vijandig en competitief wezen is dat geen middel schuwt om zijn individuele behoeften te bevredigen. Gezien deze gewelddadige aard van de mens was er volgens hem maar één politieke remedie die de burgers tegen zichzelf en tegen anderen kon beschermen: een soevereine macht (Leviathan), aldus zijn krachtdadige antwoord op de constante fysieke dreiging die zo onstuitbaar aanwezig is bij wezens behept met de neiging tot wederzijdse vernietiging.

Ruim een eeuw na de publicatie van Hobbes' *Leviathan* (1651) beweerde Jean-Jacques Rousseau exact het tegenovergestelde. Rousseau karakteriseerde de menselijke aard juist niet als conflictueus, maar als enkel behept met eigenliefde en *pitié*. Het leven werd niet gekenmerkt door sterke onafhankelijkheid maar door onderlinge afhankelijkheid, door het gedeelde universele instinct tot zelfbehoud, en een zekere intuïtie dat het individuele voortbestaan sterk verbonden was met dat van zijn medeburgers.

Overtuigd van de slechtheid van de mens keerde Hobbes zich tot het autoritarianisme, als het enige middel om de mensheid tegen zichzelf te beschermen. Naar Rousseaus stellige overtuiging waren de aannames van Hobbes echter fundamenteel onjuist en schreef hij aan van nature goedaardige wezens eigenschappen toe (zoals ijdel eigenbelang en gewelddadige prestatiedrang) die in feite waren aangeleerd. De mens, zo wierp Rousseau tegen, was geperverteerd door zijn omgeving, door onrechtmatige maatschappelijke en politieke verhoudingen die een verder onschuldige soort hadden gecorrumpeerd, en verlossing van de burgers moest dan ook worden gezocht in een hervorming van hun instituties. Waar Hobbes geweld zag als middel tot zelfbescherming en de burgers hun politieke wil ontnam, was Rousseau voorstander van actieve deelname van de bevolking via een systeem van volkssoevereiniteit gebaseerd op de democratische rechtsprincipes van vrijheid, onafhankelijkheid en gelijkheid. Rousseau verwierp de standpunten van Hobbes en beweerde dat de heersende ongelijkheid en onderdrukking die zo funest waren voor het menselijk welzijn, enkel en alleen opgeheven konden worden door een grotere betrokkenheid van het volk.

well lead us to view Rousseau's vision with a measure of scepticism. Ours is, after all, a climate of fear and instability, a period ravaged by insecurity and terrorist violence, and divided by increasing socioeconomic inequalities between wealthy and poor, technologically advanced and traditional cultures, western and 'third world' nations, Jihad fundamentalism and McWorld capitalism. We might therefore be tempted to side with Hobbes, to embrace his dour view of human nature and accept brute (and brutal) force as the best means of cultivating a safer, more harmonious globe.

Although our age is beset by chaos and confusion, one in whose publics are scattered and diffuse, terrorized by violence and surveillance alike, we must nonetheless resist the temptation to succumb to anxiety and mistrust. Siding with Rousseau, it is our opinion that democracy alone offers a strong rebuttal to both terrorism and surveillance, to the coercive, at times despotic roles these tactics press upon individuals stripped of their very capacity for self-rule.

Passive politics understood as spectatorship engenders fear. Active politics dispels fear, and in this sense engaged democracy is the only effective response to fear. The least fearful Americans after September 11, 2001 were the rescue workers at Ground Zero, exposed to great risk, but engaged in common work and hence unafraid. Likewise, the passengers on the flight brought down over Pennsylvania, because they rebelled and took back control of the airplane, must have felt less terror than the passengers on the other three flights who rode passively to their deaths. The psychological fact is that fear breeds fear and spectatorship reinforces it. Shock and awe cannot foster liberty. Only democracy as active citizenship can do that, which means that fear's natural enemy is democratic citizenship.

Building democratic communities is, however, no easy task. After all, any individual (and, indeed, nation) that wishes to secure itself from terror by forging a world of free expression needs to be at least as interested in smart citizens as it is in smart bombs or smart surveillance technology. Yet our first instincts are often

In de toestand waarin de moderne wereld verkeert, is een zekere mate van scepsis ten aanzien van de visie van Rousseau wel te verwachten. We leven immers in een klimaat van angst en instabiliteit, een tijdperk dat wordt geteisterd door onzekerheid en terroristisch geweld en waarin verdeling heerst door de toenemende sociaal-economische ongelijkheid tussen arm en rijk, tussen technologisch hoog-ontwikkelde en traditionele culturen, westerse en derdewereldlanden, *jihad*-fundamentalisme en *McWorld*-kapitalisme. We zouden dus geneigd kunnen zijn de kant van Hobbes te kiezen, zijn sombere kijk op de menselijke aard over te nemen en bruut en meedogenloos geweld te accepteren als het beste middel om een veiliger, harmonieuzer wereld te bewerkstelligen.

Hoewel het tijdperk waarin we leven wordt beheerst door chaos en verwarring, de maatschappij verdeeld en versplinterd is onder de terreur van het geweld en de beveiligingsmaatregelen in reactie daarop, moeten we de verleiding weerstaan ten prooi te vallen aan angst en wantrouwen. We scharen ons achter Rousseau in de opvatting dat alleen democratie een voldoende krachtig middel is in de strijd tegen het terrorisme en de beveiligings-maatregelen, en tegen de dwingende, heerszuchtige rol die individuen – ontdaan van hun recht op zelfbeschikking – door dergelijke tactieken opgelegd krijgen.

Passieve politiek, omstanderpolitiek, veroorzaakt angst. Actieve politiek verdrijft angst en in dat opzicht is een geëngageerde democratie de enige effectieve respons op angst. De Amerikanen die de minste angst voelden na 9/11 waren de reddings-werkers op Ground Zero. Ze waren blootgesteld aan grote risico's, maar omdat ze samen aan het werk waren, voelden ze geen angst. Zo hebben de passagiers van de vlucht die in Pennsylvania neer-stortte waarschijnlijk ook minder angst gevoeld dan de passagiers op de drie andere vluchten, die passief hun dood tegemoet gingen, omdat ze in opstand kwamen en weer de controle over het vliegtuig kregen. Het is een psychologisch gegeven dat angst slechts angst oproept en dat passief toe-kijken het gevoel van angst alleen maar versterkt. De strategie van 'shock and awe' is niet bevorder-lijk voor vrijheid en onafhankelijkheid. Die bereiken we alleen door actieve deelname van burgers aan de democratie, wat betekent dat democratisch staatsburgerschap de natuurlijke vijand is van angst.

Het is echter geen eenvoudige opgave om een actieve democratische samenleving op te bouwen. Per slot van rekening moet elk individu (en elk land) dat zichzelf wil behoeden voor terreur door een

to monitor behaviour and exaggerate difference, to train soldiers rather than build an autonomous citizenry. Such violent, paranoiac measures are short-sighted; as Mr. Smith's case demonstrates, they ultimately accentuate individual feelings of alienation and helplessness.

As we argue, Hobbes' worldview is both misleading and impractical: a politics rooted in suspicion and fear only begets fear. This phenomenon is certainly evident in the case of our beleaguered patient. Big Brother lurks through the video cameras perched above his street corners and lodged within his cash machines. Under the auspices of public safety and welfare, this technology monitors his every move on Middelland Street. Yet under the indiscriminate gaze of a video camera, every individual is reduced to the role of a potential suspect. As Smith's anxiety demonstrates, far from cultivating a strong civil society or engaged citizenship, tactical surveillance only compounds individual fear and alienation.[1]

In a time and space where democratic values of tolerance, liberty, freedom, and collective harmony are needed to unite a world ravaged by violence and insecurity, democratic nations have looked to misguided solutions – monitoring, military manoeuvres, racial profiling – in fighting an enemy which can only be bested by global civil society. The war against fear can succeed only in a world of peaceful democracies, yet this transformation will come neither easily nor quickly. We must, after all, recall that democracy's most important virtue (and condition necessary for its development) is patience. Democracy is a process not an end, and it moves in stages. The patience of this progression needs to be recognized as indispensable to the success of those who today have embarked on the difficult journey.

Democratic outcomes ultimately depend on democratic struggle and the readiness of citizens – or those who would be citizens – to wage it. Mr. Smith may consider himself unique, but he is just one in a world of individuals both discontent with current affairs and capable of enacting change. To do so he must draw a deep

wereld met vrije meningsuiting te creëren, minstens zo geïnteresseerd zijn in intelligente burgers als in intelligente bommen of intelligente beveiligings-technologie. Toch zijn we instinctief vaak geneigd gedrag in de gaten te houden en verschillen aan te dikken. We trainen liever soldaten dan autonome burgers dat we kweken. Zulke gewelddadige, paranoïde maatregelen zijn kortzichtig: het geval van Mr. Smith toont aan dat ze uiteindelijk alleen maar de individuele gevoelens van vervreemding en machteloosheid benadrukken.

Zoals we trachten aan te tonen is het wereldbeeld van Hobbes zowel misleidend als onpraktisch: een politiek van achterdocht en angst brengt slechts angst voort. Dit fenomeen is maar al te duidelijk in het voorbeeld van onze beproefde patiënt. Big Brother kijkt mee door de videocamera's die boven elke straathoek hangen en die zijn geïn-stalleerd bij geldautomaten. Onder het mom van openbare veiligheid en welzijn wordt onze patiënt door middel van deze technologie in al zijn bewe-gingen op straat gevolgd. Maar onder de aselecte blik van een videocamera wordt elk individu gereduceerd tot de rol van potentiële verdachte. Uit de angstgevoelens van Smith blijkt wel dat tactisch toezicht enkel bijdraagt aan individuele angst en vervreemding, en allesbehalve bevorder-lijk is voor een sterke samenleving of betrokken burgerschap.[1]

In een tijd en ruimte waarin democratische waar-den zoals tolerantie, onafhankelijkheid, vrijheid en collectieve harmonie nodig zijn om een wereld te verenigen die wordt geteisterd door geweld en onzekerheid, hebben de democratische landen de verkeerde middelen aangewend – toezicht, militaire manoeuvres, etnische profilering – om een vijand te bestrijden die alleen verslagen kan worden door een wereldwijde burgerlijke samenleving. De oorlog tegen angst kan alleen worden gewonnen in een wereld van vreedzame democratieën, maar een dergelijke transformatie zal niet gemakkelijk of snel verlopen. Tenslotte is, dat moeten we niet vergeten, geduld de beste eigenschap van een democratie, en een noodzakelijke voorwaarde voor haar ont-wikkeling. Democratie is geen doel op zich, maar een proces dat in fasen verloopt. Het is van belang te onderkennen dat geduld bij dit proces onont-beerlijk is, willen degenen die deze moeilijke reis vandaag de dag ondernemen, kunnen slagen.

Democratie hangt uiteindelijk af van de democra-tische strijd die ervoor is gevoerd, en van de bereidheid van de burgers (of toekomstige burgers) deze te voeren. Mr. Smith denkt wellicht dat hij de enige is, maar hij is slechts een van de velen in een

breath of courage in the face of adversity, and ask what he can do for his city, country and world alike.

After all, the empire of fear is a realm without citizens, a domain of spectators, of subjects and victims whose passivity means helplessness and whose helplessness defines and sharpens fear. Citizenship builds walls of activity around fear: this cannot prevent the doing of terrorist deeds or hegemonic surveillance, but it diminishes the psychic toll both take. Where citizens such as Smith yearn to be involved, their governments put them on the defensive; where they wish to become agents, their leaders insist their actions are relevant only when threatening or suspicious. To relinquish fear people must step out of this induced paralysis, and draw courage in the conviction that together, on local and global levels alike, they possess the power to implement change. Far from an idealist plea, this challenge reflects the hard reality of the world in which we live: an increasingly interdependent global stage which must lean to carve space for citizens to act and repair the bonds that, for better or worse, necessarily bind us to one and all alike.

Benjamin R. Barber is a American political theorist and writer. He is the Gershon and Carol Kekst Professor of Civil Society and Distinguished University Professor at the University of Maryland and a principal of Democracy Collaborative.

Joshua Karant is a Democracy Collaborative Research Fellow and Director of the CivWorld Higher Ed Initiative.

wereld vol individuen die ontevreden zijn met de gang van zaken én in staat er iets aan te doen. Hij moet zich niet uit het veld laten slaan, moed vatten en vragen wat hij kan betekenen voor zijn stad, zijn land en zijn wereld.

Want het rijk van de angst is een wereld zonder burgers, een wereld van omstanders, van onderdanen en slachtoffers wier passiviteit hen machteloos maakt, een machteloosheid die hun angst bepaalt en verscherpt. Actief burgerschap kan deze angst indammen; terroristische aanslagen of toezicht van bovenaf worden daarmee niet voorkomen, maar eisen psychisch wel een minder hoge tol. Burgers als Smith zouden dolgraag actief betrokken worden, maar de overheid dwingt hen in een defensieve rol; ze zouden graag in actie komen, maar hun leiders beweren dat hun acties alleen belang hebben als ze bedreigend of verdacht zijn. Om los te komen van de angst moeten mensen deze aan hen opgelegde verlamming van zich afschudden en moed verzamelen, in de overtuiging dat ze samen, op lokaal en op mondiaal niveau, in staat zijn dingen te veranderen. Een zware opdracht, maar zeker geen idealistisch pleidooi; eerder een reflectie van de harde werkelijkheid waarin we leven: een mondiaal toneel waar men in toenemende mate onderling afhankelijk is van elkaar is en waar ruimte gemaakt moet worden voor burgers om te handelen en de banden te herstellen die ons – in voor- en tegenspoed – met alle anderen verbinden.

Benjamin Barber is een Amerikaans politicoloog en schrijver. Hij is Gershon & Carol Kekst professor Civil Society, professor aan de universiteit van Maryland en directeur van Democracy Collaborative.

Joshua Karant is als onderzoeker verbonden aan Democracy Collaborative en directeur van het CivWorld Higher Ed Initiative.

1   Such tactics may shock Smith, yet they can hardly be considered revolutionary. In September, 2003 the Chicago police force added 250 cameras to the 2,000 already installed throughout the downtown metropolitan area. A program inspired by the United Kingdom's use of video technology to fight Irish terrorists, it now incorporates innovations developed by Las Vegas casinos to immediately alert authorities of any deviant activity. As justification Mayor Richard M. Daley argued that 'Cameras are the equivalent of hundreds of sets of eyes. They're the next best thing to having police officers stationed at every potential trouble spot.'
Stephen Kinzer, 'Chicago Moving to "Smart" Surveillance Cameras', The New York Times, September 21, 2004.

1   Dergelijke tactieken zijn wellicht schokkend voor Smith, maar ze zijn nauwelijks revolutionair te noemen. In september 2003 voegde de politie van Chicago nog eens 250 camera's toe aan de 2000 die al overal in het zakencentrum van de stad waren geïnstalleerd. Dit vormde onderdeel van een programma dat was geïnspireerd op het gebruik van videotechnologie in het Verenigd Koninkrijk om Ierse terroristen te bestrijden, en omvat nu innovaties die ontwikkeld zijn door de casino's in Las Vegas om de autoriteiten direct attent te maken op elke afwijkende activiteit. Burgemeester van Chicago Richard M. Daley voerde als rechtvaardiging aan dat 'camera's het equivalent zijn van honderden paren ogen. Ze zijn de beste oplossing als je niet op elke potentiële risicoplek politieagenten wilt neerzetten.'
Stephen Kinzer, 'Chicago Moving to "Smart" Surveillance Cameras', *The New York Times*, 21 september 2004.

WARM MILK
WITH HONEY

The investigative design of DUS centres on making fear into an object of discourse. Their methodology shows parallels to the Freudian talking cure. Just as the psychoanalyst has techniques at his disposal to extract meaningful details from a patient, DUS uses design techniques, a 'designing cure', to extract the desired information from its environment.

From the idea that a solution to the problem of fear and space is an impossible task, DUS formulates a strategy in three phases. In the first phase, information is collected about the context through an acknowledgement of fear. Conditions, environments, architectures and circumstances are researched and analysed so that the contours of the problem become visible. In the second phase, DUS employs investigative design to make the area's possibilities clear and discussible within these contours. A diagnosis is made, as it were, of the environment and its use, functioning and interpretation. The DUS team deciphers both the spatial model as well as the way in which users operate within it. In the third and final phase, a priority list (as a consensus, greatest common denominator, or ideology) is drawn up of a number of phenomena that occasion fear. These are then translated into a number of interventions within the existing context. The interventions attempt both to meet the wishes and needs of the target group and to combine this with the DUS philosophy – in this case the acknowledgement of fear instead of its denial.

Both literally and figuratively, DUS is in an 'in-between situation' – between *top-down* and *bottom-up* – which entails filtering the appropriate alternative from among all the possible options.

WARME MELK
MET HONING

In het onderzoekend ontwerp van DUS staat het bespreekbaar maken van angst centraal. Hun benaderingswijze vertoont verwantschap met de Freudiaanse *talking cure*. Zoals de psychoanaliticus technieken ter beschikking heeft om zinvolle gegevens aan een patiënt te ontfutselen, zo doet DUS een beroep op ontwerptechnieken, *designing cure*, om de gewenste informatie uit zijn omgeving te halen. Vertrekkend vanuit het idee dat het oplossen van het probleem van angst en ruimte een onmogelijke opgave is, stelt DUS een strategie voor in drie fases. In de eerste fase wordt door erkenning van angst informatie verzameld over de context. Condities, omgevingen, architecturen en omstandigheden worden onderzocht en geanalyseerd tot de contouren van het probleem zichtbaar worden. In de tweede fase zet DUS het onderzoekend ontwerp in om binnen deze contouren de moeilijkheden van het gebied duidelijk en bespreekbaar te maken.

Er wordt als het ware een diagnose gesteld van de omgeving en het gebruik, het functioneren en de interpretatie ervan. Het DUS-team leest zowel de ruimtelijke vormgeving als de manier waarop de gebruikers zich daarbinnen bewegen. In de derde fase ten slotte wordt een prioriteitenlijst (als consensus, grootste gemene deler of ideologie) opgesteld van een aantal fenomenen die aanleiding geven tot angst. Deze worden op hun beurt vertaald in een aantal ingrepen binnen de bestaande context. De ingrepen pogen én aansluiting te vinden bij de wensen en noden van de doelgroep én die te combineren met de filosofie van DUS, in dit geval: de erkenning van angst in plaats van de ontkenning ervan.

DUS bevindt zich zowel letterlijk als figuurlijk in een 'tussenin-situatie', tussen *top-down* en *bottom-up*, daarbij gaat het erom dat uit alle mogelijke alternatieven de geschiktste wordt gefilterd.

WARME MELK MET HONING

WARM MILK WITH HONEY

PROJECT:
DUS
BIJDRAGE / CONTRIBUTION:
MARK PIMLOTT

Een bescheiden uitwisseling
Mark Pimlott

Op het thema van *Groepsportretten* – Angst en Ruimte – en het heersende idee dat onze steden plaatsen van angst zijn geworden, reageert de architectengroep DUS met de gedachte dat mensen troost moet worden geboden, zoals een moeder voor haar kind zorgt en het 'melk en honing' geeft. Deze huis-tuin-en-keukenstrategie lijkt vreemd aan de architectuur, althans strijdig met haar conventies, die in het moderne tijdperk primair zijn gekenschetst in termen van rationalisatie en de bijbehorende afstandelijkheid. Hoe moeten we de doelstelling en strategie van DUS dan interpreteren? Als DUS (Martine de Wit, Hans Vermeulen en Hedwig Heinsman, architecten, voor dit project aangevuld met Roel Spits, onderwijzer, Arja Boon, organisatie/pr-specialist en Diana Kuip, journaliste en columniste) geen club van maatschappelijk werkers is, hoe moeten we dan hun strategie en het mogelijke succes daarvan duiden in de context van de creatieve praktijken die in de samenleving werkzaam zijn?
Ten eerste is er de kwestie van de 'angst'. Het is van belang te constateren dat de angst zoals die momenteel in onze steden wordt ervaren slechts gedeeltelijk te maken heeft met een realiteit van escalerend geweld. Geweld is in de steden altijd voorgekomen; toch is duidelijk dat de door de mensen ervaren 'angst' direct gerelateerd is aan een acuut en toenemend gevoel van vervreemding. Deze vervreemding, die diametraal staat tegenover het eveneens bestaande, bevrijdende gevoel van anonimiteit dat een stad oproept, vloeit voort uit het karakter van de kapitalistische, moderne en modernistische stad, maar is er ook onderdeel van. Ze is een product van de Verlichting en het bourgeoisstreven naar zelfontplooiing. We zijn getuige van voortdurende wordings- en vernietigingsprocessen en van een constant toenemende mobiliteit en leegstand op plaatselijke, regionale en mondiale schaal.
De voortdurende transformaties zijn voelbaar in gebouwen, omgevingen en gemeenschappen. Veroudering, overtolligheid en verval grijpen om zich heen.

Straight to the Point

**What are you afraid of? We asked around and came up with the following top three:
1. traffic accidents 2. muggers 3. diseases**

**And rightly so.
However, it was a different fear that started us thinking. One person hesitantly admitted being afraid of sharks.Yes, sharks. The lady in question never felt safe in or on the water.**

So, is this an irrelevant fear? It looks like it at first. At least, when we calculate the odds. After all, what is the chance of a stray shark taking a tasty bite out of this lady's calf off the coast of Scheveningen? Pretty small, we would say. And the probability that a shark enthusiast sees a good opportunity to treat his pet to a day's paddling? Minimal. But does this mean the lady's fear is completely unjustified? Can we guarantee for a full hundred percent that this will never, really never ever actually happen to her? No, we definitely can't do that. Er... help? H'm. But, actually, there is also something reassuring about this. Because if we can conclude that we can NEVER guarantee safety, we can say that EVERY fear is justified. So? Well, we then automatically stop making a value judgement, and consider the fear of sharks to be as justified as the fear of a traffic accident.

Warm Milk with Honey

For a while it may have looked like it, but the DUS team by no means intended to focus on the shark issue. But we do take the recognition of EVERY fear as our basis. Because if a world without risks does not exist, we need not try to create one.

So the next question is: can we imagine anything that might mitigate fear? Is it possible to design something that can set something positive against any possible fear? Something, in fact, that offers us comfort? Because as far as we're concerned that is the key word. So our mission became a quest for a design that contains a comfort factor. A positive addition to an environment. Or, as DUS soothingly called it: a cup of warm milk with honey, in the public space.

And here is the result. What you have in your hands is the Double-Thick-DUS-Do part. Play with it and find your COMFORT. Have lots of fun. Love, DUS.

Met de deur in huis

Waar ben je bang voor? We gingen het gewoon eens
vragen, wat ons de volgende topdrie opleverde
1. Verkeersongeval 2. Overvallers 3. Ziekten

Ja, terecht.
Toch was het een andere angst die ons aan het
denken zette. Eén persoon gaf namelijk
schoorvoetend toe bang te zijn voor haaien.
Inderdaad, haaien.
Te water waande mevrouw zich nergens veilig.

Is dit nou een irrelevante angst? Zo lijkt het in eerste
instantie wel. Tenminste, als we een kansberekening
toepassen. Want hoe groot is de kans dat een
verdwaalde haai voor de kust van Scheveningen in
de kuit van de mevrouw een smakelijk hapje ziet?
Klein lijkt ons. En de waarschijnlijkheid dat
een haaihobbyist zijn kans schoon ziet om zijn
troeteldier op een dagje Tikibad te trakteren?
Miniem. Maar is de angst van mevrouw dan volstrekt
onterecht? Kunnen we haar voor de volle honderd
procent garanderen dat dit nooit, maar dan ook echt
nooit zal gebeuren? Nou nee, dat niet. Eh… help?
Tja. Maar eigenlijk heeft het ook wel iets
rustgevends. Want als we kunnen concluderen
dat we nóóit veiligheid kunnen garanderen,
kunnen we stellen dat élke angst gegrond is. Dus?
Nou, dan houden we automatisch op een waarde-
oordeel te vellen, en vinden we de angst voor haaien
net zo terecht als angst voor een verkeersongeval.

Warme melk met honing

Het mocht wellicht even zo lijken, het team van
DUS was geenszins van plan zich te gaan richten
op haaienproblematiek. Maar het erkennen
van élke angst nemen we wel als basis.
Want als een wereld zonder risico's niet bestaat,
hoeven we ook niet te trachten deze te scheppen.
De volgende vraag luidt dan: is er iets denkbaar wat
angst wellicht zou kunnen verzachten?
Is het mogelijk iets te ontwerpen wat iets positiefs –
tegenover welke angst dan ook – kan stellen?
Iets wat troost biedt dus eigenlijk? Want daar ligt wat ons
betreft het sleutelwoord. Onze missie werd dan ook
een zoektocht naar een ontwerp dat een troostende
factor in zicht draagt. Een positieve toevoeging aan
een omgeving. Of, zoals DUS het liefkozendnoemde:
een kopje warme melk met honing,
in de openbare ruimte.

Zie hier het resultaat. Voor je ligt het
Dubbel-Dikke-DUS-Doedeel. Speel mee en vind je troost.
Veel plezier. Kus, DUS

Het modernisme haakte in op dit ken-
merk van de moderne tijd (dat al een
rol speelt sinds de Industriële Revolutie)
en het uitvergroot, ten eerste door de
chaotische vitaliteit ervan naar voren
te halen en die vitaliteit vervolgens te
vergelijken met de werking van een
machine.
In architectonische zin moest de stad
getransformeerd worden tot een beeld
van hoe moderniteit werkte: geordend,
machinaal, emotieloos. 'Tuer la rue'
betekende dat de plek waar de menselijke interactie in de stad traditioneel
plaatsvond – de straat – een negatieve
betekenis kreeg en moest worden geëli-
mineerd. Hiervoor in de plaats moesten
snelwegen, parken en collectieve
gebouwen komen die in hun indeling
en rangschikking expliciet uitdrukking
moesten geven aan de impliciete hiër-
archieën van de samenleving, altijd
in verband met werk, productie en
consumptie. Er mocht geen intimiteit
zijn, geen 'plek', geen identificatie met
iets anders dan een collectief, monu-
mentaal menselijk project. Het individu
zou gewoon een functioneel onderdeel
zijn van de machinerie en tevreden met
die rol. De emblematische programma's
uit de heroïsche periode van de moderne architectuur, zoals Hilberseimers
Grossstadtarchitektur of Le Corbusiers
Ville contemporaine, waren, zij het
alleen in symbolische zin, representatief voor de krachten die binnen de
moderne kapitalistische samenlevingen
werkzaam waren. Ze waren niet te
verwezenlijken door één enkele entiteit.
De programma's en theorieën waarmee
ze gepaard gingen, waren niet alleen
politiek en economisch implausibel; ze
gingen regelrecht in tegen de menselijke
manier van leven. Hoe onuitvoerbaar
deze theorieën en voorstellen ook leken,
de avant-garde bleef ze in periodieke
congressen (CIAM) gewoon ontwikkelen
en na 1945 konden ze alsnog worden
verwezenlijkt. In de projecten voor de
herbouw van Europese steden werd de
theorie tot orthodoxie. De armzaligheid
van de modernistische visie op de stad
is inmiddels overduidelijk: de resultaten
zijn illustratief voor het feitelijke gebrek
aan respect voor individuen en hun
mogelijke manieren van samenleven.
En al is het modernisme tegenwoordig

COMFORT

ENGLISH
1 WARM MILK WITH HONEY
2 CALLING THE POLICE
3 IN BED WITH MAMMY AND DADDY
4 GIFTS

NEDERLANDS
1 WARME MELK MET HONING
2 POLITIE HALEN
3 BIJ PAPA EN MAMA IN BED
4 CADEAUTJES

SCARY PEOPLE / ENGE MENSEN

1 BIN LADEN
2 FLASHER / POTLOODVENTER
3 MAGICA DE SPELL / ZWARTE MAGICA
4 MICHAEL JACKSON

SCARY NICE THINGS

ENGLISH
1 HAUNTED HOUSE
2 SIX FLAGS HOLLAND
3 FLYING
4 MOVING

NEDERLANDS
1 SPOOKHUIS
2 SIX FLAGS HOLLAND
3 VLIEGEN
4 VERHUIZEN

SCARY THINGS ON TELEVISION

ENGLISH
1 THRILLER
2 CNN
3 CHILDBIRTH
4 WAR

NEDERLANDS
1 GRIEZELFILM
2 JOURNAAL
3 BEVALLINGEN
4 OORLOG

SCARY THINGS AT SCHOOL

ENGLISH
1 REPORT CARD
2 GYMNASTIC TEACHER
3 RUTHLESS CLASSMATES
4 SHOW AND TELL

NEDERLANDS
1 SCHOOLRAPPORT
2 GYMLERARES
3 MEEDOGENLOZE KLASGENOOTJES
4 SPREEKBEURT

SCARY THINGS IN YOUR BED

ENGLISH
1 ALLIGATOR UNDER YOUR BED
2 GHOSTS BEHIND THE WINDOW
3 PHANTOM AT THE ATTIC
4 CLOTH-MONSTER

NEDERLANDS
1 KROKODIL ONDER JE BED
2 SPOKEN ACHTER HET RAAM
3 KLOPGEEST OP ZOLDER
4 KLERENMONSTER

SCARY THINGS ON HOLIDAY

ENGLISH
1 SNAKES AND SPIDERS
2 NASTY THINGS IN YOUR DINER
3 SHARKS
4 DIRTY TOILETS

NEDERLANDS
1 SLANGEN EN SPINNEN
2 VIEZE DINGEN IN JE ETEN
3 HAAIEN
4 VIEZE TOILETTEN

SCARY THINGS ON THE STREET

ENGLISH
1 (BLACK) DOGS
2 DRUNKEN PEOPLE
3 TRAFFIC ACCIDENT
4 DARKNESS

NEDERLANDS
1 (ZWARTE) HONDEN
2 DRONKEN MENSEN
3 VERKEERSONGELUK
4 DONKER

wat minder openlijk aanwezig, de denk-
wereld bestaat nog. Waar eens de
ideologie regeerde, heerst nu de markt.
De mythe dat het individu baas is over
zijn eigen leven wordt gepropageerd
via de belofte van bevrijding door
consumptie: niet alleen kan men door
objecten te kopen zijn wensen vervullen,
men benadrukt ook zijn individualiteit en
onderscheidt zich van anderen. Nergens
is zo duidelijk zichtbaar waar een op
vrije kapitaalvorming gebaseerde stedelijke ontwikkeling toe leidt als in
de Verenigde Staten, hoewel Europa
zich in zijn streven naar terugdringing
van het staatsapparaat ten gunste van
de 'individuele vrijheid' steeds meer aan
dit voorbeeld spiegelt. Concurrentie
tussen individuen wordt aangemoedigd,
wat hen verder van elkaar isoleert,
terwijl ze toch leden zijn van dezelfde
consumentenstam. Individuen worden
geacht hun eigen weg te zoeken in een
zinloze, vijandige wereld, afgescheiden
van anderen, geatomiseerd.
Consumptie wordt aangeprezen met
de belofte van 'bescherming' tegen
de wereld en tegen angst. Deze angst
wordt door de staat en het bedrijfsleven
(via de media) uitgespeeld als een
constante dreiging die alleen door de
staat en het bedrijfsleven kan worden
beheerst, wat de angst alleen maar
vergroot. Om zich veilig te weten moet
men geloven in de overheid en in de
markt, die zich sterk maken voor indivi-
dualisering en consumptie. Het publiek
krijgt te horen dat er een strijd gaande
is, en dat wij (nu weer als groep, omdat
dat voor dit soort oproep beter uitkomt)
waakzaam moeten zijn, dat we ons
moeten beschermen tegen elke vorm
van rebellie, van binnenuit en van
buitenaf. Deze retoriek van de overheid
en het bedrijfsleven brengt een pani-
sche omwenteling van angst teweeg,
waarbij het individu steeds op voet van
oorlog verkeert met de ander. Maar wie
is die ander? Als men het niet voortdurend eens is over de voortdurend
veranderende regels voor de strijd, dan
is de ander, de vijand, het individu zelf.
De strijd maakt mensen onverschillig
voor het lot van anderen, waardoor de
idee van een gemeenschap moeilijk in
stand kan worden gehouden. Wie hoort
erbij en wie niet? We horen er geen van

Performance

**Before we start designing, we first of all want
to get to know about characters. Those of both
the location and its inhabitants. Questionnaires
and participatory discussions fall short in this
respect because above all we want to SEE.
That's why we're starting with a performance.
Or call it an intervention, or a test case,
or a bombardment. We surprise and explain nothing.
That's the only way to get a pure reaction.
We arrange a living room,
occupy a square, or hijack a building.
And then you come and look.
You're certainly welcome.
Fancy a coffee?
Want to sit near the window?
Draw something yourself?
Want a biscuit?**

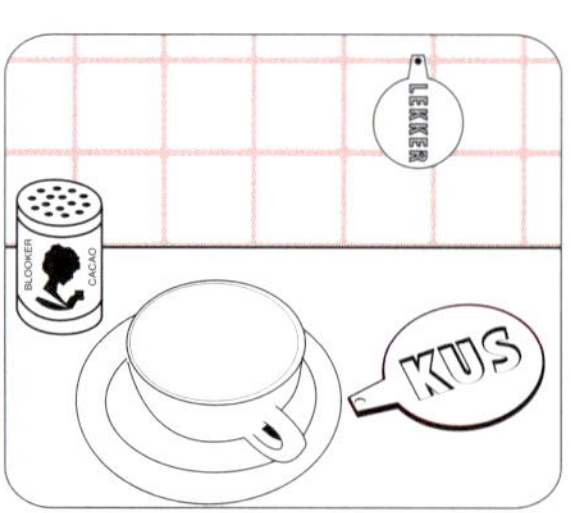

Performance

**Voor we gaan ontwerpen, willen we allereerst
karakters leren kennen. Van zowel de plek
als de bewoners ervan. Vragenlijstjes en
inspraakrondes schieten daarin dus te kort,
want we willen het vooral zien.
Daarom starten met een performance.
Of noem het een interventie, of een testcase of een
bombardement. We verrassen, en leggen niets uit.
Alleen zo krijg je een pure reactie.
We richten een huiskamer in,
bezetten een plein of kapen een pand.
En kom dan maar kijken.
Wees vooral welkom.
Wat heb je in je koffie?
Zit je graag bij het raam?
Teken zelf eens.
Wil je een koekje?**

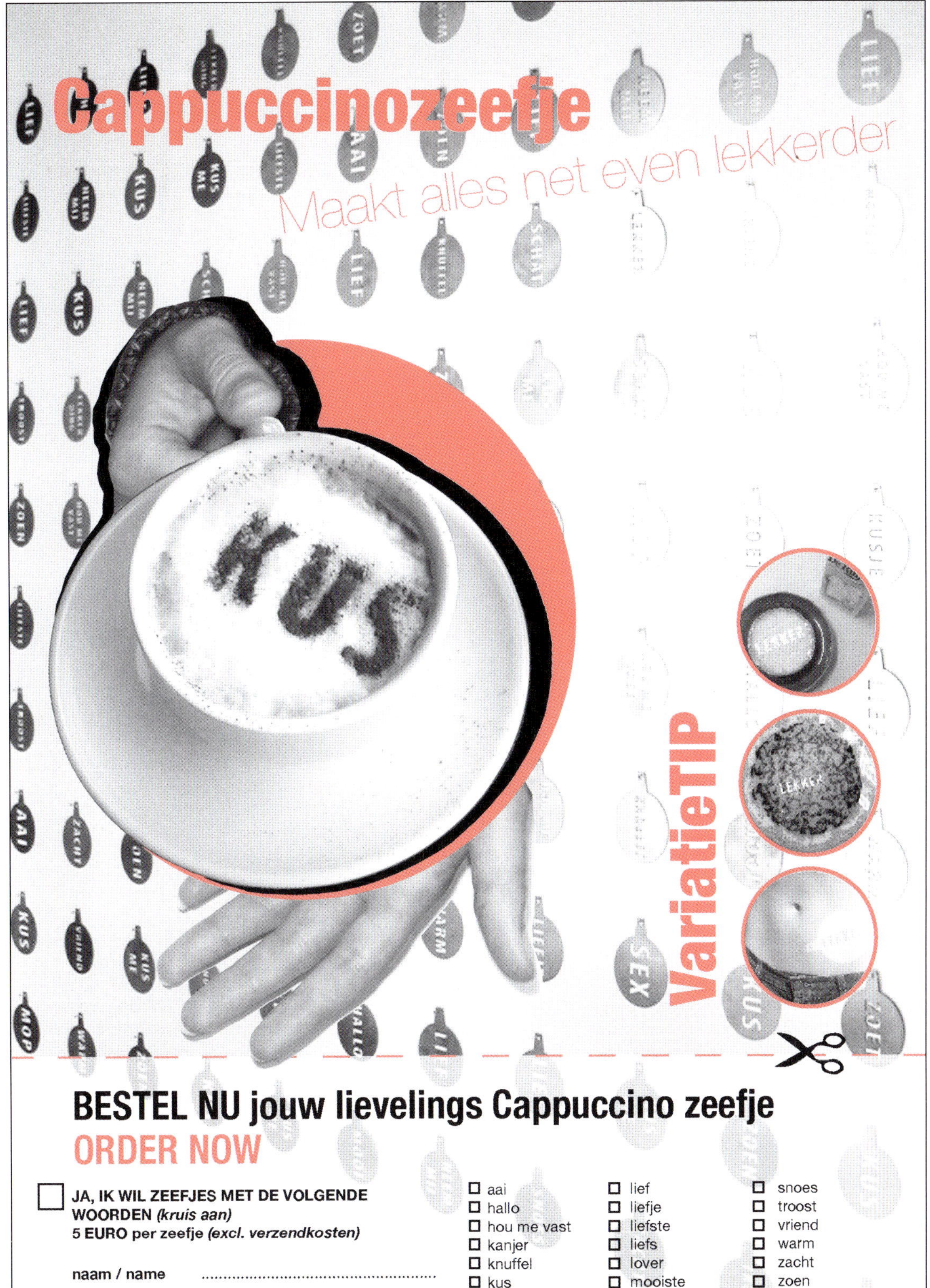
Cappuccinozeefje
Maakt alles net even lekkerder
KUS
VariatieTIP
LEKKER
BESTEL NU jouw lievelings Cappuccino zeefje
ORDER NOW
JA, IK WIL ZEEFJES MET DE VOLGENDE
WOORDEN (kruis aan)
5 EURO per zeefje (excl. verzendkosten)
naam / name
adres / address
postcode / stad
email
aai
hallo
hou me vast
kanjer
knuffel
kus
kusje
kus me
lekker
lekker ding
lief
liefje
liefste
liefs
lover
mooiste
mop
neem mij
schat
sex
snoes
troost
vriend
warm
zacht
zoen
zoet

allen bij, tenzij we ons aanpassen, tenzij we consumeren.

De strategie van DUS in het project 'Angst en Ruimte', dat geheel in lijn is met de rest van hun praktijk tot nu toe, is niet louter een balsem voor de wonden van de gekwetsten in een samenleving die uit winnaars en verliezers bestaat. Ze heeft veel meer te maken met sociale relaties. Als lid van de samenleving wil DUS iets 'schenken', of het nu gaat om het bieden van een simpel onderkomen (Cocoon), vrijheid in het uiten van persoonlijke wensen bij hun ontwerpen van woningen (Urban Interiors – Stedelijke Interieurs), een plaats waar kinderen kunnen spelen en waar het karakter van die plaats door die kinderen zelf wordt ontwikkeld (Blackboard House – Krijthuisje), eten maken (Prora Bäckerei, Bakje Troost) of ontspanningsruimten voor plaatselijke bewoners (Porches – Veranda's). Hiermee staan ze aan de architectuurpraktijk verbonden privileges als auteurschap en zeggenschap af, en bieden in plaats daarvan troost: die troost komt niet noodzakelijkerwijs van hun 'melk en honing', maar van de zeer reële uitwisselingen en relaties – de ontwikkeling van een gemeenschap – die door hun 'geschenken' mogelijk worden gemaakt. In de lange Noord-Europese traditie van maatschappelijk georiënteerde praktijken waarin architectuur een rol speelt, hebben gemeenschapsbewegingen en collectieven allemaal geprobeerd een systeem binnen het systeem te creëren, om de hardheid van de vrije markt buiten de deur te houden en de impact van de meedogenloze en onverbiddelijke ontwikkeling van de stad in de moderne tijd te verzachten. Deze bewegingen omhelzen echter vaak een utopisme dat uiteindelijk regressief is, omdat het de idee van ontwikkeling verwerpt. In deze geprojecteerde utopia's wordt de wereld voorgesteld als stilstaand, bevroren in eenstemmigheid. Het werk van DUS is niet utopisch. Integendeel, verandering wordt toegejuicht. DUS heeft geen vooropgestelde ideeën over de resultaten of hoe die eruit moeten zien. Ze vinden dat wat ze doen betekenis moet hebben voor de mensen voor wie en met wie ze werken.

House of Chalk

**What is comfort? We want to find out from the pupils of the Paulusschool in Bos en Lommer.
But how do you put 8 to 10 year-olds so much at ease that they dare to talk about something as intimate as this? In response to this question, DUS built a little interview house in blackboard black, and called it the House of Chalk.
One day this house just appears in the classroom. Roel, the teacher, doesn't say anything about it. The children immediately take over this new spot by going inside it to read. A day later they find the chalk. Transformation. The house changed colour twenty times, children colour their signatures and in this way appropriate it for themselves.
A Golden Choice.
Because this little house itself already turns out to be an example of offering comfort by spatial means. In the classroom it offers the children an opportunity to create a place of their own.
To use the chalk is to conquer.**

Krijthuisje

**Wat is troost?
Dit willen we weten van leerlingen aan de Paulusschool in Bos en Lommer.
Maar hoe stel je kinderen in de leeftijd van 8 tot 10 jaar dermate op hun gemak dat ze over zoiets intiems durven te praten?
DUS bouwt als antwoord daarop een schoolbordzwart interviewhuisje, oftewel het Krijthuisje.
Op een dag staat het huisje zomaar in de klas. Meester Roel vertelt er verder niets bij.
Al meteen veroveren de kinderen deze nieuwe plek door er in te gaan lezen. Een dag later wordt het krijt gevonden. Transformatie.
Het huisje verandert (twin)tig keer van kleur, kinderen kleuren hun handtekening en eigenen zich het op die manier toe.
Een Gouden Greep.
Want dit huisje zélf blijkt al een voorbeeld van het ruimtelijk bieden van troost. Binnen het klaslokaal biedt het kinderen een kans een eigen plekje te creëren.
Krijten is overwinnen.**

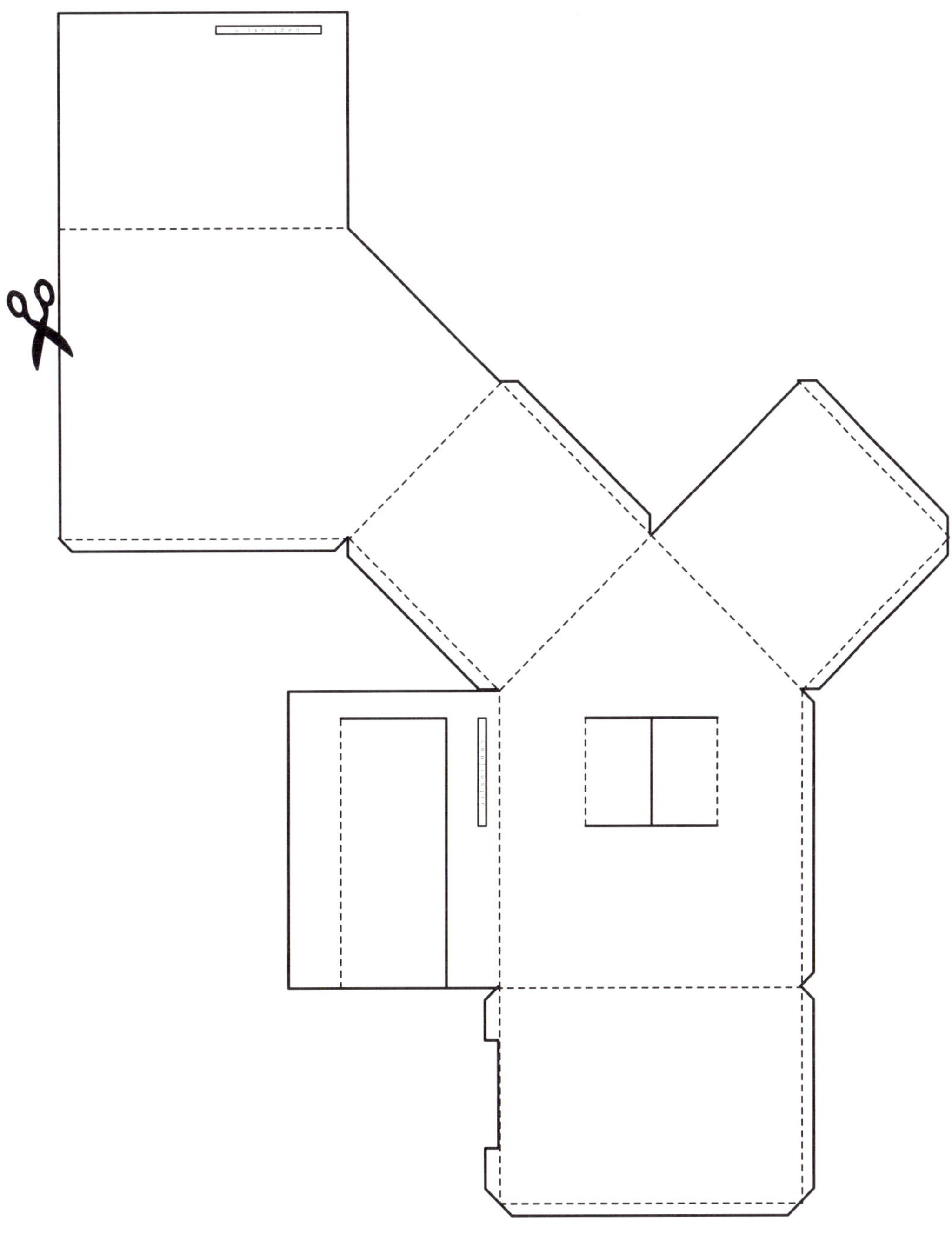

Krijthuisje, Paulusschool, Bos en Lommer, Amsterdam (oplossing zie p. 83)
House of Chalk, Paulusschool, Bos en Lommer, Amsterdam (solution on p. 83)

Ze vinden dat niet alleen zijzelf die
betekenis moeten bepalen, maar dat
deze moet voortkomen uit de uitwisseling tussen hen en de mensen, tussen
hun gebouwde werk en de manier
waarop dit wordt gebruikt en geïnterpreteerd; een uitwisseling die voortdurend in beweging is, voortdurend
verandert. DUS biedt de mensen eten,
dingen, situaties en plaatsen, in veel

DUS, schemerlamptafel, 2004

gevallen als 'geschenk'. Door dit te
doen krijgen ze een beter inzicht in
de plaatsen waar ze hun werk maken,
en zo wordt hun werk een vorm van
analyse. Het proces van informatie verzamelen en uitwisselen is consequent
open en ondogmatisch. Zowel in hun
directe omgang als in hun gebouwde
werk streven ze naar een ongedwong-
en communicatie met mensen. Ze
houden de mogelijkheid open dat
andere zaken dan hun eigen wensen
een plaats krijgen in hun werk en dat
ze het resultaat beïnvloeden.
Het gebouwde werk dat ze voorstellen
is vaak ambigu, zowel qua vorm als
qua mogelijk gebruik; het is vaak tijdelijk of provisorisch. In die ambiguïteit
speelt de status van het werk een centrale rol. Is het publiek of privé? Hun
werk toont aan dat dergelijke definities
niet onwrikbaar zijn, maar veranderlijk.
Bakje Troost (2004), een door DUS ontworpen café waar cappuccino wordt
geschonken (met liefdevolle, leuke of
troostrijke woorden in cacaopoeder
erop gestrooid) werd de plaatselijke
bevolking aangeboden midden op
een plein in Katendrecht (Rotterdam).
Er werd een gordijn over de straat
gespannen en het café verscheen als
bij toverslag op het plein. Het anders
lege plein werd bruikbaar, huiselijk
en zichtbaar. De bezoekers konden

zichzelf zien als een tijdelijke gemeenschap die in het openbaar met elkaar
praatte in plaats van als geïsoleerde
individuen. DUS hield vóór het gordijn
een soort optreden en initieerde zo een
sociaal ruilcontract: zij zorgden voor de
koffie en vervolgens spraken de mensen
met elkaar. Zo ontstaan samenlevingen.
In het door DUS voorgestelde *Veranda's*
(2004) wil het team bescheiden toe-
voegingen creëren voor gebouwen in
Katendrecht. Zo zou er een overdekte
bank kunnen worden toegevoegd
aan de winkelpui van de plaatselijke
tatoeagesalon. De bank kan door de
buurtbewoners worden gebruikt om
gewoon wat te zitten of door de klan-
ten van Tattoo Bob die op hun beurt
wachten. Het wordt opzettelijk in het
ongewisse gehouden of de bank een
openbare toevoeging is aan de studio
van Tattoo Bob of een privé-toevoeging
die door het publiek wordt gebruikt. Het
werk is ambigu maar communicatief en
bevordert de intimiteit op straat, de
ruimte van het publiek. Deze privé-
activiteit is ook voor iedereen zichtbaar.
Is deze plek van Tattoo Bob, van de
straat, of van de mensen die er zitten
te praten? En waarom zitten ze daar te
praten? Omdat het kan, en dat is de
subtiliteit die deel wordt van het werk
van DUS, en van hoe dat werk werkt.
Het werk van DUS is van een schaal
waarop men de werking ervan nog kan
zien, zodat men uitwisselingen tussen
individuen ziet en geen imaginaire
effecten op abstracte gebruikers. De
mechanismen van de werken zijn niet
onzichtbaar, zoals bij veel architectuur
of bij het instrumentele beleid van de
overheid het geval is. In dit opzicht en
op deze schaal doet hun werk denken
aan dat van kunstenaars die zich bezighouden met architectuur en het publiek,
met name Dan Graham en Gordon
Matta-Clark. Het is de maatvoering van
de door deze kunstenaars gecreëerde
werken en situaties waartegen het
werk van DUS kan worden afgezet.
Met Dan Graham deelt DUS de ambitie
constructies en situaties te maken in
de publieke ruimte, die qua functie
ambivalent en ambigu zijn. Ook bij deze
werken is het doel de relaties tussen
mensen 'neer te zetten' en zo een
kritisch bewustzijn van zichzelf en de

Schemerlamptafel (oplossing zie p. 83)
Meeting station (solution on p. 83)

Dan Graham, *Interior Design for Space Showing Videotapes*, 1986-87

ander te creëren. De gebruikers of ontvangers van hun werk – dat niet zozeer wordt opgelegd als wel aangeboden – ervaren zichzelf als individuen tussen anderen, zich van zichzelf bewust als actieve spelers in de openbare wereld. Hun werk staat zelfs nog dichter bij dat van Gordon Matta-Clark, wiens voorstellen te beschouwen zijn als gekant tegen de architectuur, met generositeit en onthulling als centrale impuls. Het werk van DUS kan net als het zijne worden beschreven als een soort geschenk: feesten, koken, bijna-architectuur/bijna-gebeurtenissen. De overeenkomsten met *Food* (1971) en werken als *Fresh Air Cart* (1972) zijn treffend. De mensen herinneren zich deze werken omdat ze hun ervaring van de wereld en van andere mensen hebben veranderd. Dit is belangrijk. De acties van DUS helpen de gemeen-

Gordon Matta-Clark, *Food*, 1971

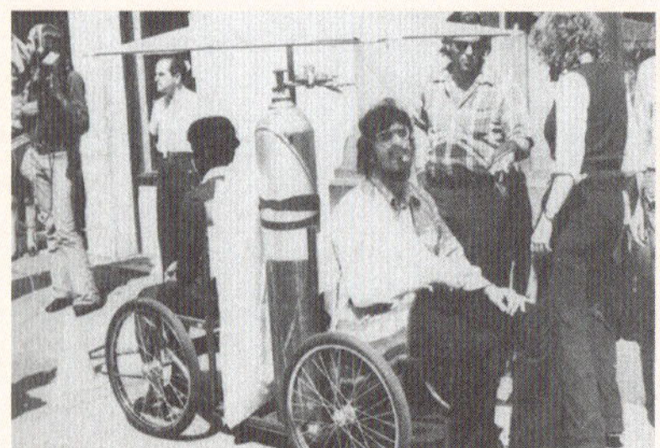
Gordon Matta-Clark, *Fresh Air Cart*, 1972

Warme melk met honing   DUS   Mark Pimlott   65

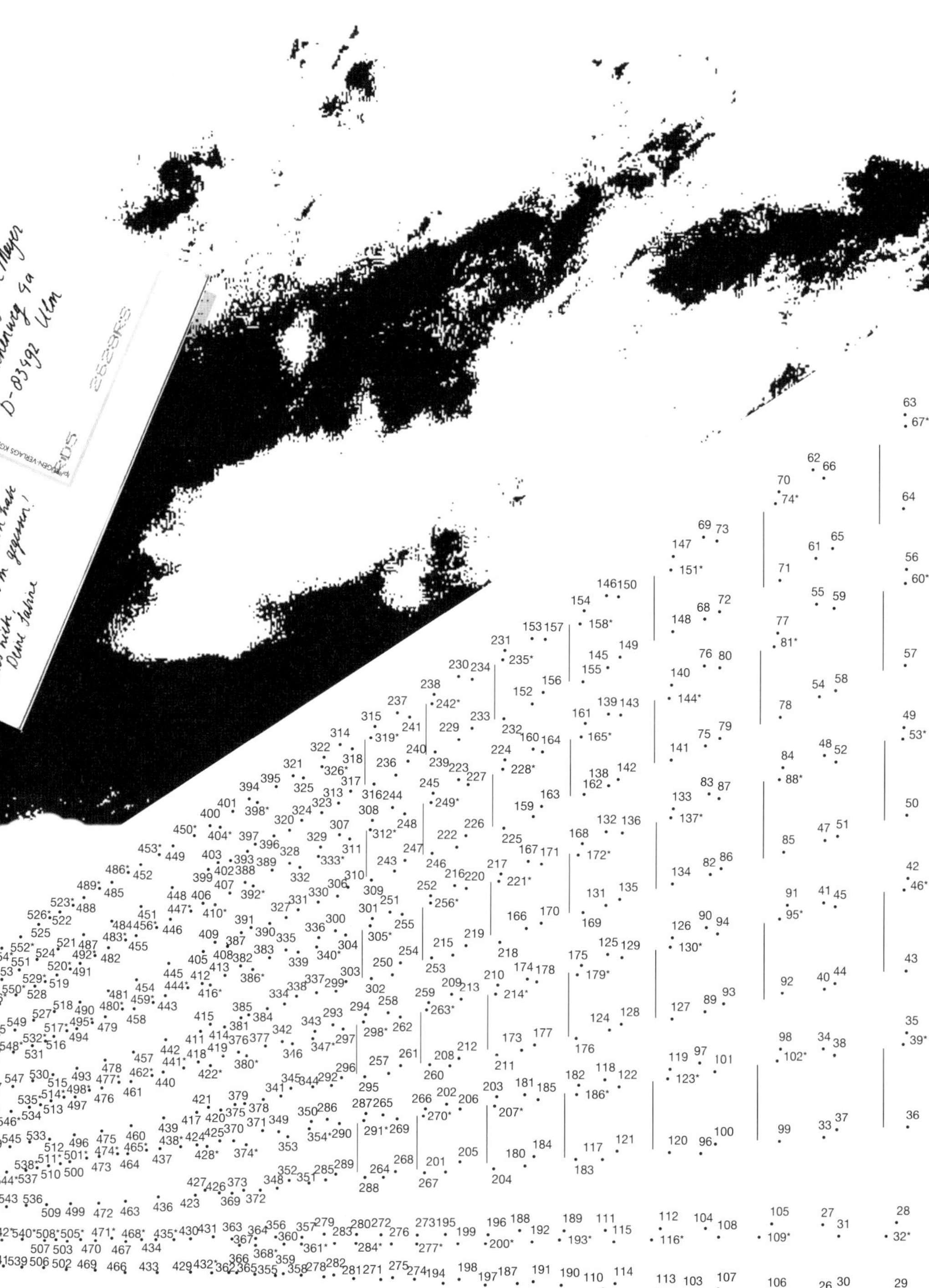

Prora, Rügen, Duitsland, `haal je pen van het papier en ga verder bij het volgende nummer (oplossing zie p. 83)
Prora, Rügen, Germany, `take your pen off the paper and go on to the next number (solution on p. 83)

schap zichzelf als zodanig te zien; wat hun werk effectief maakt, is niet alleen hoe het eruitziet, maar ook dat het in je geheugen blijft hangen. De bescheiden voorstellen van DUS, die noch architectuur, noch stedenbouw, noch programma lijken te zijn – allemaal preoccupaties van de huidige neo-avant-garde – zijn werkzaam op het niveau van het menselijk bewustzijn. Door hun werk te maken in het ambigue gebied tussen intieme of privé-ruimte (en het daar geldende ruilmiddel, communicatie) en publieke ruimte (die zichzelf niet als zodanig bewust is), maken ze werk dat op een manier tegelijkertijd beide en geen van beide is. Ze openen de mogelijkheden voor communicatie tussen het privé- en het publieke domein, tussen individuen. Mensen onthouden dat soort dingen.

Mark Pimlott is kunstenaar en ontwerper in Den Haag en Londen. Op dit moment is hij hoogleraar met betrekking tot de praktijk aan de afdeling Architectuur (Interieur) aan de Technische Universiteit Delft.

Nibble, Nibble Like a Mouse

But then.
We put the house outside.
Is it still your domain, we asked the children.
Can the house of chalk stand in an open space too? No, say the children, I wouldn't go inside.
After all, who would it belong to?
The public space seems to be a no-man's-land.
And everyone is an outlaw.
Inside is safe, outside isn't.
The private domain lies behind the façades.
That's where we live and laugh.
Outside, it's no-man's-land,
empty and without identity.
With a hard but thin dividing line.
Indeed.
A door.

Knibbelknabbelknuisje

Maar dan.
We zetten het huisje buiten.
Is het nu nog jullie domein, vragen we de kinderen.
Kan het krijthuisje ook op een plein?
Nee, zeggen de kinderen, dan ga ik er niet in.
Want van wie is het?
Het publieke lijkt een niemandsland.
En iedereen is vogelvrij.
Binnen is het veilig, buiten niet.
Achter de gevels schuilt het privé.
Daar word geleefd en gelachen.
Daarbuiten is het niemandsland,
leeg en identiteitsloos.
Met een harde, maar dunne scheidslijn.
Inderdaad.
Een deur.

Krijthuisje, Paulusschool, Bos en Lommer, Amsterdam (oplossing zie p. 83)
House of Chalk, Paulusschool, Bos en Lommer, Amsterdam (solution on p. 83)

A Modest Exchange
Mark Pimlott

In response to the theme of *Group Portraits* – Fear and Space – and the prevailing idea that our cities have become places of fear, the architecture group DUS have responded with the notion that people need to be given comfort, as a mother might take care of her child, administering 'milk and honey'. This home-spun and old-fashioned strategy seems non-architectural, or at least runs against the conventions of architecture, which have been characterized in modernity as primarily directed towards rationalization and its distance. So, how might this objective and strategy of DUS be seen? If DUS (Martine de Wit, Hans Vermeulen and Hedwig Heinsman, architects, joined for this project by Roel Spits, an elementary school teacher, Arja Boon, an organization /public relations specialist and Diana Kuip, a journalist and columnist) are not social workers, then how can one understand and evaluate their strategy and its possible success in the context of creative practices working within society?

First, there is the issue of 'fear'. It is important to understand that fear as experienced at present within our cities is only partly to do with the reality of escalating violence. Cities have always hosted violence, yet it is clear that the popular experience of 'fear' is closely tied to an acute and gathering sense of alienation. This alienation, quite opposite to the parallel, liberating sense of anonymity that cities encourage, is both a consequence and part of the nature of the city of capitalism, modernity and Modernism. It is a product of the Enlightenment and the bourgeois drive to selfrealization. We are witnesses to continual processes of generation and destruction and to everincreasing mobility and abandonment at local, regional and global scales. The constant transformations are felt in buildings, environments and communities. There is an abundance of obsolescence, redundancy, ruin. Modernism latched onto this characteristic of modernity (which had been present since the arrival of industrialization) and exaggerated it, first, by marking impressions of its chaotic vitality and then likening that vitality to the workings of machinery. In architecture, the city was to be transformed into an image of the operations of modernity: ordered, machine-like, without sentiment. Tuer la rue meant that the traditional place of human interaction in the city – the street – acquired negative significance and had to be eliminated. In its place were to be motorways, parks and collective buildings whose arrangement and organization would explicitly reflect the implicit hierarchies of society,all bound to work, production and consumption. There would be no intimacy, no place, no identification with anything other than a collective, monumental human project. The individual would be simply a working part of the machinery and happy to be so. The emblematic schemes of modern architecture's heroic period, such as Hilberseimer's Großstadtarchitektur or Le Corbusier's Ville contemporaine, were representative yet only symbolic of the forces at work in modern capitalist societies. No single entity could possibly organize their realization. The schemes and the theories that accompanied them were implausible not only on political and economic levels: they were utterly at odds with the ways of people. Yet the unrealizable nature of these proposals and theories, which continued to be developed by the avant-garde in regular congresses (CIAM), became realizable after 1945. In the projects for rebuilding European cities, theory became orthodoxy. The poverty of the Modernist urban vision is now all too evident: the results are emblematic of the actual and limited regard for individuals and their possible associations. And though the overt forms of Modernism may have now faded, its world remains. Where there was ideology, there is now the market. The myth of the individual's power to control his life is propagated by the promise of liberation through consumption: not only can one fulfil one's desires, but one can guarantee one's individuality, one's difference through the acquisition of objects. Nowhere

Das längste und leckerste Gebäude der Welt

Prora Kekse

Prora Bäckerei
NEU
SEIT 2003

Zutaten: Mehl, Zucker, Butter, Salz, ohne Konservierungsmittel.

Ofenfrisch verpackt:

Nach dem Öffnen schnell alle Kekse formieren und genießen, genau so wie bei Oma.

100 % Kekse

Inhalt: 11 Kekse
Maßstab 1 : 500
Mit 11 Keksen lässt sich ein Gebäude block von Prora zusammen setzen.
Die ganze Anlage besteht aus 8 Blöcken!

Ab 2005 im Sortiment der Bäckerei Peters in Sassnitz auf Rügen / D

Beste Kauf

currently is the fruit of urban development based on the free generation of
capital more clearly seen than in the
United States, which Europe – in its
drive to diminish the apparatus of state
in favour of 'individual freedom' – seeks
to emulate. Competition is enforced
between individuals, increasing their
isolation from each other, even while
they are part of the same consuming
tribe. Individuals are expected to
make their own way in a senseless,
hostile world, dissociated from others,
atomized. Consumption promises
'protection' from the world and from
fear, which is played on by state and
business (through media) as a constant
presence that only state and business
have the power to control, which fear
enhances. To be safe, one has to believe in government and the market,
who advocate individuation and consumption. The public is told there is a
battle going on, we (a group again, for
the purposes of this kind of address)
are asked to be vigilant, to protect
ourselves against insurgency of all
kinds, from within and without. This
rhetoric of government and business
elicits a panic of constant revolution,
where the individual is always at war
with the other. But who is the other?
If one is not in constant agreement
with the constantly changing terms of
engagement, then the other, the enemy
is the individual himself. The battle
yields an indifference to the fortunes of
others, rendering the idea of community
difficult to sustain. Who is in and who is
out? We are all out, unless we conform,
unless we consume.

DUS's strategy in the project Fear and
Space, consistent with the entirety
of their practice to date, is not merely
some sort of balm on the wounds of
the injured in a society where there are
winners and losers. Rather, it is closely
associated with social relationships. As
members of society, DUS give, whether
they are offering simple shelter (Cocoon);
freedom in the expression of personal
desires in their designs for dwellings
(Urban Interiors); a place for children
to play, where the character of the
place is developed by those children
(Blackboard House); making food

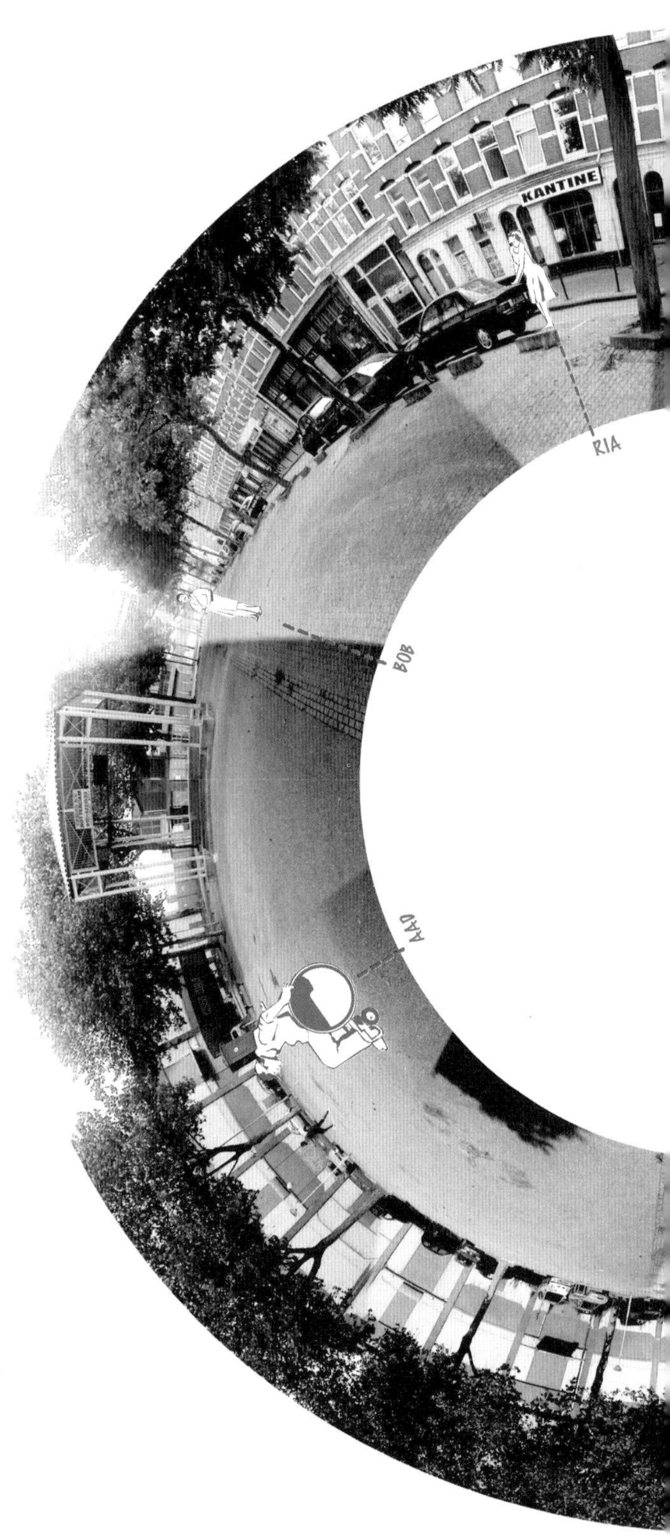

Deliplein, Katendrecht, Rotterdam
Deliplein, Katendrecht, Rotterdam

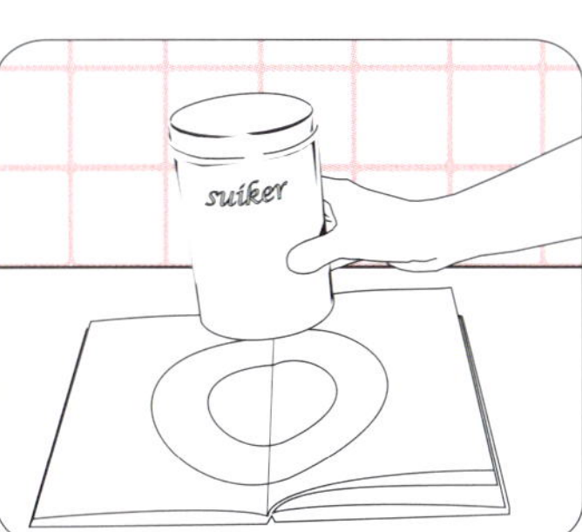
suiker

suiker

(Prora Bäckerei, Bakje Troost) or donating public relaxation spaces for locals (Porches). In doing so, they surrender the privilege of authorship and imposition that comes with the practice of architecture, and instead offer comfort: the comfort does not necessarily come from their 'milk and honey', but from the very real exchanges and relationships – the generation of community – made possible by their 'gifts'. In the long-standing tradition of socially-oriented practices involving architecture in northern Europe, community move-ments and cooperatives have all attempted to form systems within the system, trying to keep the harshness of the free market at bay and cushion the impact of the ruthless and unstoppable development of the city in modernity. Yet, the utopianism these movements often espouse is finally regressive because they reject development. For these projected Utopias, the world is seen as standing still, frozen in agreement.
DUS's work is not Utopian. Rather, it embraces change. DUS does not have predetermined ideas about either out-comes or appearances. They feel that what they do should have meaning specific to the people they work for and with. They think that this meaning should not come only from them, but from the exchange that takes place between themselves and people, be-tween their built work and the way it is used and interpreted: an exchange

DUS, meeting station, 2004

that is always in flux, always changing. DUS offers food, things, situations and places to people, frequently as 'gifts'. The performance of these acts leads to understandings of the places where

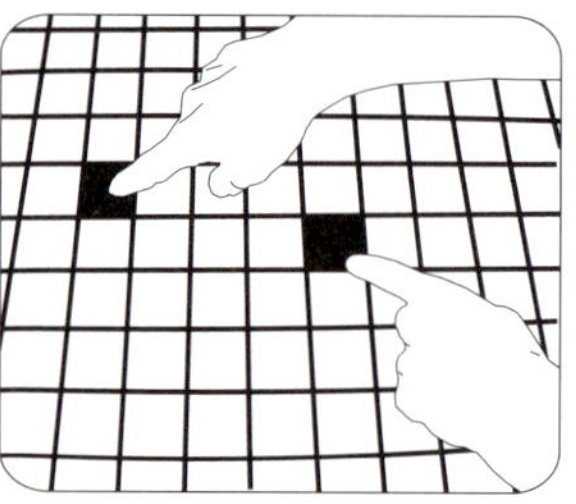

Bakje Troost Bar memory (oplossing p. 83)
'Cup of Comfort' Bar memory (solution on p. 83)

Warme melk met honing DUS   Mark Pimlott

they make their work, and becomes a kind of analysis. Their processes of gathering information and exchange are consequently open and undogmatic. They propose easy communication with people, both directly and in their built work. They allow the possibility of things other than their own desires to enter their work, to influence its outcome. The built work they propose is often ambiguous, both in its forms and in its possible uses; it is often temporary or improvisational. Particularly central to its ambiguity is its status. Is it public or private? Their work demonstrates that these definitions are not fixed, but fluid. Bakje Troost (2004), a café dispensing cappuccino (with powdered cocoa exhortations of love, pleasure and comfort) made by DUS was offered to local people in the middle of a square in Katendrecht (Rotterdam). A curtain was hung across the road and the café appeared, as if by magic, in the square. Normally empty, the square became usable, homely and visible. Visitors were able to understand themselves as a temporary community, speaking to each other in public, rather than as isolated individuals. DUS put on a kind of performance in front of the curtain and thus initiated a social contract of exchange: they provided the coffee, and then people spoke to each other. This is how societies come into being. In their proposed Porches (2004), DUS plans to make modest additions to buildings in Katendrecht. For example, a covered bench is to be added to the local Tattoo parlour's shop front. The bench is intended to be used by local people to hang out and also by customers waiting to be customized by Tattoo Bob. It is intentionally unclear whether their work is a public addition to Tattoo Bob's studio or a private addition being used by the public. It is ambiguous but communicative, encouraging intimacy in the street, the space of the public. This private activity will also be on display. Does this place belong to Tattoo Bob, or to the street, or to the people talking there? And why are they talking there? Because they can, and it is this subtlety which becomes part of DUS's work, the work that DUS's work does. DUS's work is still at a scale where it

The 'Cup of Comfort' Bar

**The door is intriguing.
Only four centimetres between private and public. Between safety and danger.
We want to tone down this particular transition.
This led to the 'Cup that Cheers' bar
on the Deliplein in Rotterdam.
An ordinary bar.
An ordinary square.
But different.
Out of joint, because we connect
inside and outside, and thereby make a place.
Inside out, as simple as a curtain.
Coupling instead of separating.
16 metres thick.
Private becomes public, or doesn't it?
The result: confrontation,
laughing people and coffee.
A surprise.**

Bakje Troost Bar

**De deur intrigeert.
4 centimeter maar, tussen privé en publiek.
Tussen veilig en gevaar.
Deze specifieke overgang willen we verzachten.
Daarom een Bakje Troost Bar,
op het Deliplein in Rotterdam.
Een gewone bar.
Een gewoon plein.
Maar dan anders.
Uit z'n voegen, want we verbinden
binnen met buiten, en maken zo een plek.
Binnenstebuiten, zo simpel als een gordijn.
Een koppeling in plaats van een scheiding.
Dikte 16 meter.
Privé wordt publiek of toch niet?
Gevolg: confrontatie, lachende mensen en koffie.
Een verrassing.**

Warme melk met honing   DUS   Mark Pimlott

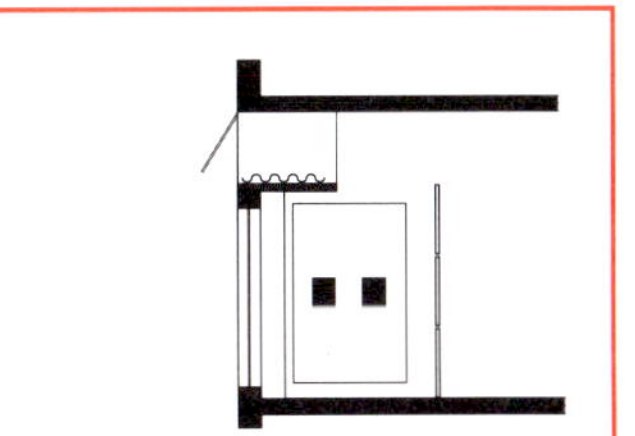

Bakje Troost Bar, Deliplein, Rotterdam
'Cup of Comfort' Bar, Deliplein, Rotterdam

is possible to see its workings, to see exchanges between individuals rather than imagining effects upon abstract users. The mechanisms of the works are not invisible, as is the case with much architecture, or the instrumental policies of government. In this way and at this scale, their work resembles that of artists involved with architecture and the public, particularly Dan Graham and Gordon Matta-Clark. It is against the measure of works and situations generated by these artists that DUS's work can be situated. With regard to

Dan Graham, *Interior Design for Space Showing Videotapes*, 1986-87

Dan Graham, DUS shares the ambition of making constructions and situations in the space of the public that are functionally ambivalent or ambiguous. The purpose of these works is similarly to expose the scene of relations between people in order to generate a critical consciousness of self and other. Users or recipients of their work – offered rather than imposed – experience themselves as individuals among others, conscious of themselves as active agents in the public world. DUS's work shares still closer associations with that of Gordon Matta-Clark, whose propositions may be seen as being against architecture, having generosity and revelation as their

Gordon Matta-Clark, *Food*, 1971

Permanent?

**It is above all the lightness of the performances that we want to see in the permanent architecture we create. Like the house of chalk or the bar,
but without interventions.
Is it possible? What happens if we ourselves are not there? Can we hand it over just like that so that the user takes over the performance?**

Permanent?

**Vooral de lichtvoetigheid van de performances willen we terugzien in de permanente architectuur die we maken. Zoals het krijthuisje of de bar, maar dan zonder ingrijpen.
Kan dat? Wat gebeurt er als we er zelf niet bij zijn?
Kunnen we het zo overhandigen opdat de gebruiker de performance overneemt?**

Warme melk met honing   DUS   Mark Pimlott   77

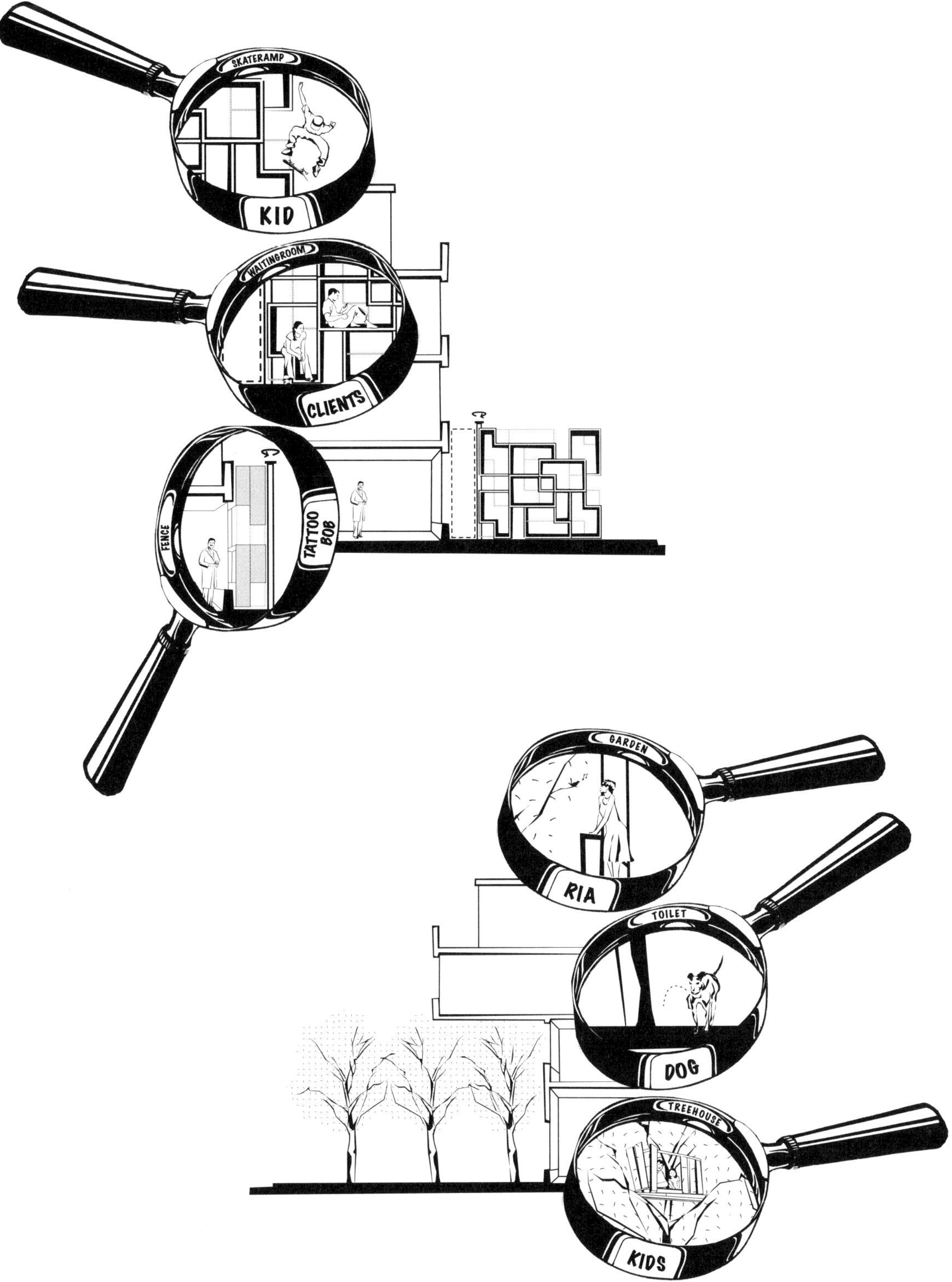

Ik zie, ik zie wat jij niet ziet 04 + 03, Deliplein, Rotterdam
'I spy with my little eye' 04 + 03, Deliplein, Rotterdam

central impulses. DUS's work, like his, can be described as a kind of gift: parties, cooking, almost-architecture/almost-events. Resemblances to *Food* (1971) and works like *Fresh Air Cart /* (1972) are striking. People remember

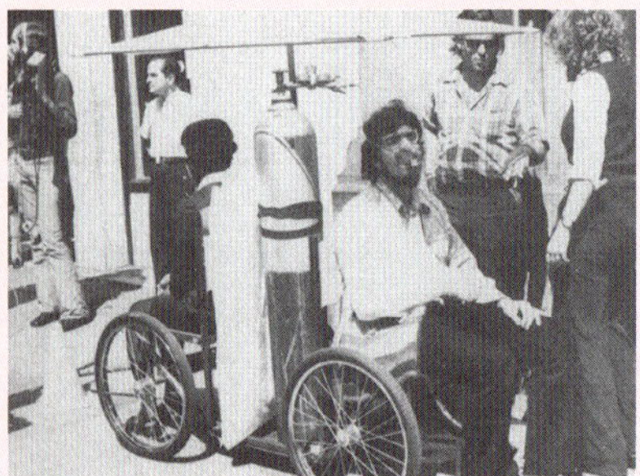

Gordon Matta-Clark, *Fresh Air Cart*, 1972

those works because they altered their experience of the world and of other people. This is important. DUS's actions help form a community's sense of itself as such; the effect of their work resides not solely in its appearance but its persisting presence in memory.
The modest propositions of DUS, which seem to be neither architecture nor urbanism nor programme – all preoccupations of the current neo-avant garde – operate at the level of human consciousness. By making their work in the ambiguous terrain between intimate or private space (and its currency of exchange, communication) and public space (which is not aware of itself as such), they make work that is both and neither. They open the possibilities for communication between the private and the public, between individuals. People remember things like that.

Mark Pimlott is an artist and designer in The Hague and London. He is currently Professor in relation to practice in the department of Architecture (Interior) at Delft University of Technology.

Strategy

**For us the challenge lies in designs
that are permanent and static but at the
same time remain changeable.
Designs that stimulate people and challenge
them to familiarize themselves with them.
The space between public and private
is perfectly suited to this. It is a domain
in its own right. A confusing area where
connections can be made and new places born.
It is precisely the confusion that is positive.
After all, is it yours? Is it mine?
May I take it or shall we share it?
From a no-man's-land to a place that summons
us to play, laugh and make human contact.
A series of verandas on the façades of the
Deliplein. Surprising, the inside comes outside,
the outside looks in. How do I have to see it?
Whose is it? Is it public, is it private?
I just go and sit there. Hello Ria (my neighbour).
Warm milk with honey?**

**The door is always open.
Go and look for yourself.
Have fun!**

Strategie

**Voor ons ligt de uitdaging in ontwerpen
die permanent en statisch zijn,
maar tegelijkertijd veranderlijk blijven.
Ontwerpen die mensen prikkelen en
uitdagen om ze zich eigen te maken.
De ruimte tussen publiek en privé leent zich
daar uitstekend voor. Het is een domein op zichzelf.
Een verwarrend gebied waar verbindingen kunnen
worden gelegd en nieuwe plekken worden geboren.
Juist die verwarring is het positieve element.
Want is het van jou? Is het van mij?
Mag ik het veroveren of zullen we delen?
Van een niemandsland tot een plek die oproept
tot spelen, lachen en menselijk contact.
Een serie veranda's aan de gevels van het Deliplein.
Verrassing, binnen komt naar buiten,
buiten kijkt naar binnen. Wat kan ik ermee?
Van wie is het? Is het publiek, is het privé?
Ik ga er gewoon zitten. Dag buurvrouw Ria.
Warme melk met honing?**

**De deur staat altijd open.
Ga zelf kijken.
Veel plezier!**

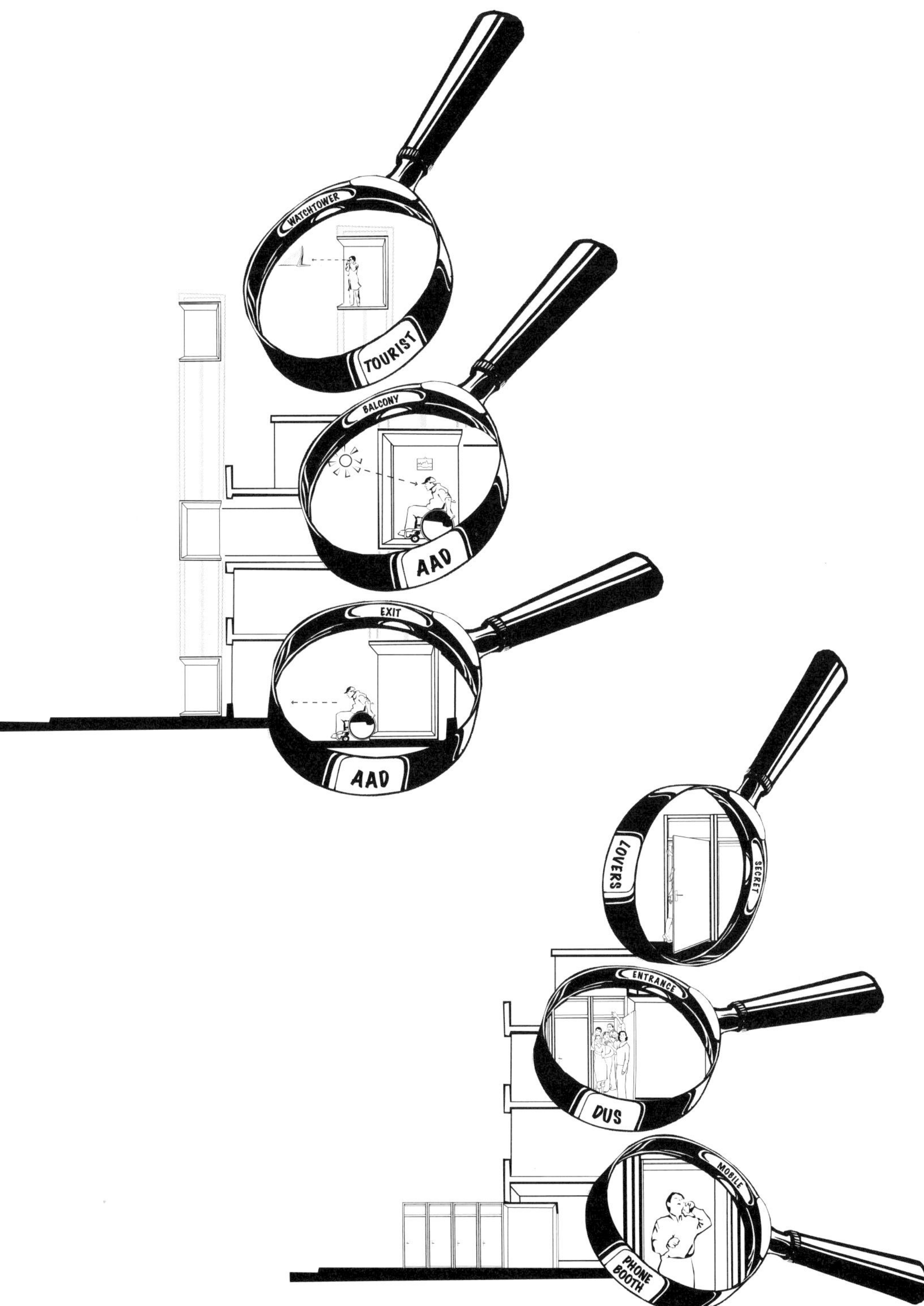

Ik zie, ik zie wat jij niet ziet 02 + 01, Deliplein, Rotterdam
'I spy with my little eye' 02 + 01, Deliplein, Rotterdam

Warm Milk with Honey   DUS   80

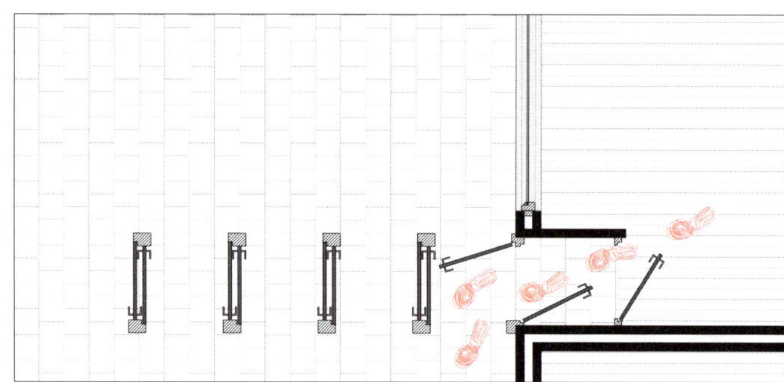

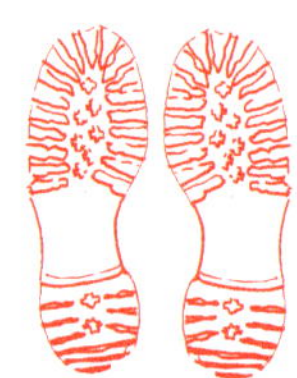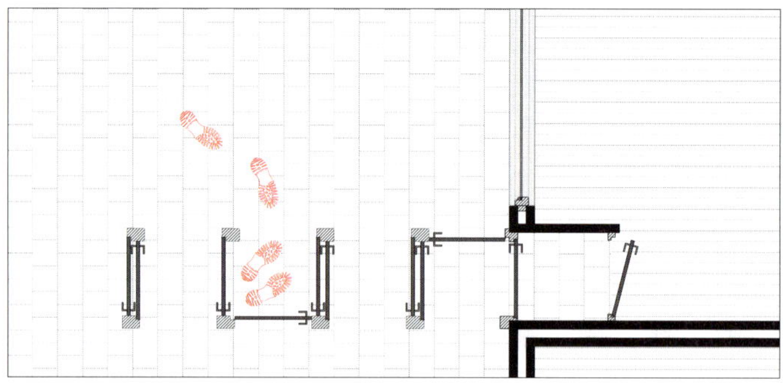

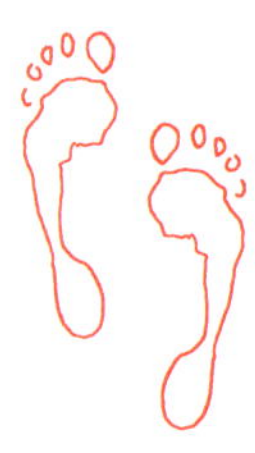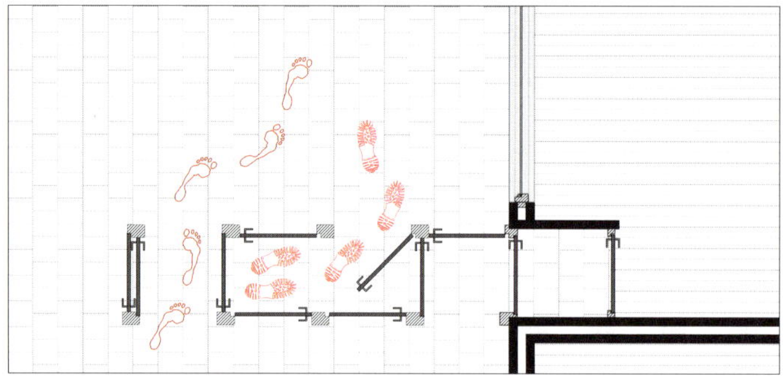

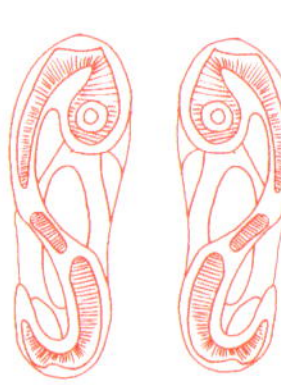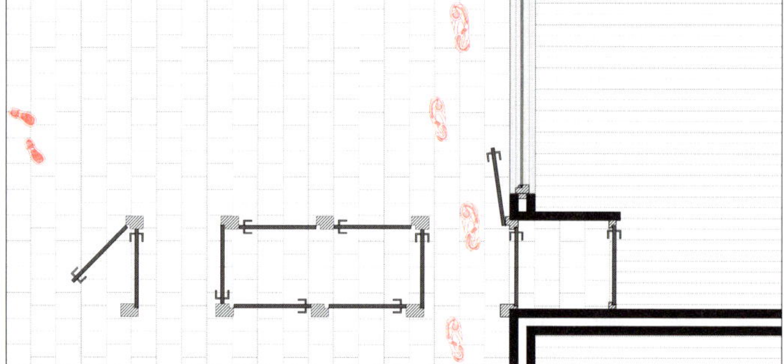

Veranda 01, Deliplein, Rotterdam
Veranda 01, Deliplein, Rotterdam

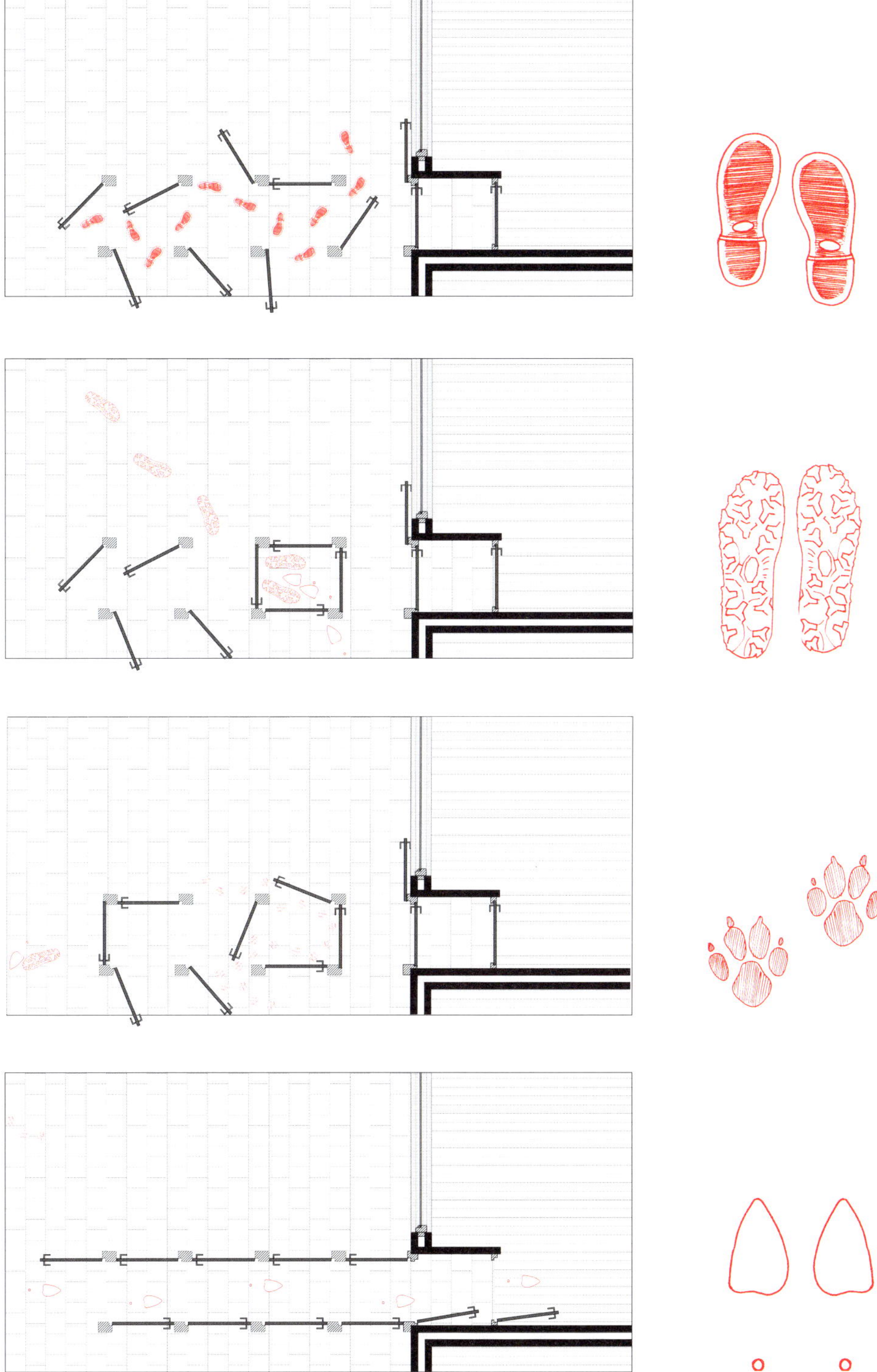

Veranda 01, Deliplein, Rotterdam
Veranda 01, Deliplein, Rotterdam

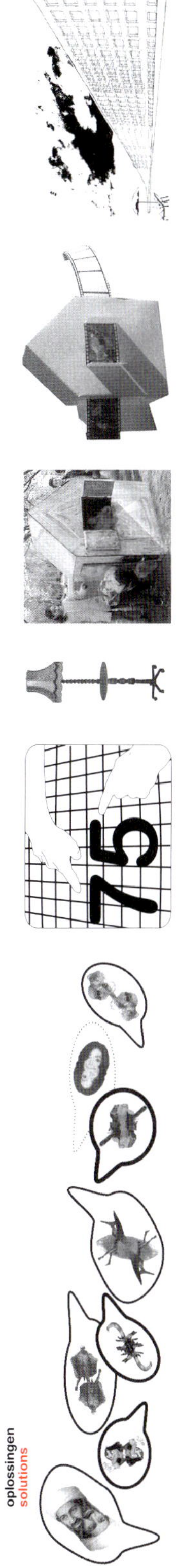

oplossingen
solutions

FEAR OF THE CITY

With 'Fear of the City' Untitled investigates what remains of the public domain, the city and its 'urbanness'. What significance can a spatial project still have? Can it still have an impact within a condition that is increasingly turning into an accumulation of suburban fragments? From a historical perspective, Untitled analyses the evolution of the urban space as a coordinating structure into an urban space as padding between suburban elements. The shift from city to architecture, from public to private and from meaning to market value plays an important role in this evolution. To Untitled, Rotterdam is the ultimate representation of a condition which places looking like a city above being a city as well as broaching the debate on the relationship between private and public space in a succinct way. Analysis of the Rotterdam cityscape extracts the hidden longings out of the city and its architecture and presents them to the city – not as a playfully composed mirror image or as a missed opportunity, but as a pivot, as a possible way to make use of what exists. This working method is a clear contradiction of the universality of the urban model, and Untitled in fact does not so much implement the illustration or representation of a method or a model as propose the concrete application of a philosophy, the result of in situ therapy. With *De stad is niet meer, leve de stad* ('The city is no more, long live the city') Untitled transcends the nihilism of confrontational opposites and eagerly faces the confrontation in its design.

ANGST VOOR DE STAD

Met 'Angst voor de stad' onderzoekt Untitled wat nog rest van het publieke domein, de stad en haar stedelijkheid. Wat kan een ruimtelijk project nog betekenen? Kan het nog impact hebben binnen een conditie die meer en meer tot een accumulatie van suburbane fragmenten verwordt? Vertrekkend vanuit een historisch perspectief analyseert Untitled de evolutie van de stedelijke ruimte als overkoepelende structuur naar een stedelijke ruimte als opvulling tussen suburbane elementen. De verschuiving van stad naar architectuur, van publiek naar privaat en van betekenis naar marktwaarde speelt een belangrijke rol in deze evolutie. Voor Untitled is Rotterdam de ultieme verbeelding van een conditie die én de schijn van de stad boven het zijn van de stad plaatst én het debat over de relatie tussen private en publieke ruimte pregnant aansnijdt. Via analyse van het Rotterdamse stadsbeeld wordt het verborgen verlangen uit de stad en haar architectuur gehaald en aan diezelfde stad gepresenteerd. Niet als een ludiek samengesteld spiegelbeeld of als gemiste kans, maar als scharnier, als mogelijkheid om met het bestaande aan de slag te gaan. Net die manier van werken spreekt de universaliteit van het stedelijke model tegen. Untitled werkt dan ook niet zozeer de illustratie of verbeelding van een methode of een model uit, maar draagt de concrete toepassing van een gedachtegoed aan, het resultaat van *in situ*-therapie. Met *De stad is niet meer, leve de stad* overstijgt Untitled het nihilisme van confronterende tegenstellingen en is het hongerig om ontwerpend de confrontatie aan te gaan.

ANGST VOOR DE STAD

FEAR OF THE CITY

PROJECT:
UNTITLED
BIJDRAGE / CONTRIBUTION:
MORITZ KÜNG

FEAR AS AN ASSOCIATIVE SPACE
Moritz Küng

1 1

Both fear (a feeling) and space (a fact) are rather broad notions. Fear is an unease, an emotional feeling that is not clearly definable, which is not experienced by everyone in the same way and under the same conditions. By contrast, space is a concrete fact and has a variety of definitions and contexts: space is cosmic, geometric, virtual, psychological, transcendental and architectural as well as historical, social, ambulant, private or public. The German phenomenologist and architect Franz Xaver Bayer devoted an entire collection of essays to the definition of spatiality[1], in which he describes 45 aspects of spatial arrangements under three headings: unlived, lived and expansive space. It appears from this that feeling and fact are not to be separated and they allow countless associations precisely because they are based on a subjective experience.

But to what extent are the two notions from the set thematic assignment actually related to each other, how can they be made to relate and what conclusions can an architect attach to them? After all, the architect's discipline means he tends to optimize spaces; and if a moment of fear occurs he will suppress it rather than make it explicitly visible.

A possible approach is to call architecture 'dangerous' or 'frightening' by definition. After all, because of his personal views of the surroundings or its future functioning, the architect imposes a particular way of thinking or an ideology on the collective (whether requested or not). Architecture always stands literally in the way and is an obstacle. Architecture of great public importance (a railway station, a courthouse) will be accepted with far fewer problems than architecture for a purely private purpose. In addition to its functionality, it is also its aesthetic form that does or does not increase public acceptance. Architecture with a high symbolic value can reinforce identity. One only has to think of the Eiffel Tower, which was built in 1887 for the centenary of the French Revolution, or more recently the Guggenheim Museum in Bilbao, which was able to inject a new economic impetus into an entire region. The greater architecture's 'common sense', its recognizability and individuality, the more society identifies with it and appropriates it.

But a contextual shift may mean that architecture is suddenly and unintentionally experienced as negative. What was once praised as a form of living that pointed the way to the future (the Bijlmer in Amsterdam, for example) may as a result of changing needs decline into a ghetto. And that which once represented technological progress, a prosperous economy and political power (think of the Twin Towers in New York) is now a symbol of impotence and fear.

But what arouses the feeling of fear? The alien and unknown? The uncontrollable? The monumental? The ambiguous? Can an image, a n event, a space or a symbol alone arouse certain fears? An arbitrary assemblage of associative series of illustrations makes an attempt to find out:

– The exorbitant monumentality of Luis Baragan's 70-m-high sculpture Faro del Comercio (1984) in Monterey/Nuevo Léon in Mexico, a double, red-painted concrete slab, which as an abstract geometric form is in extreme contrast with the landscape;

– A snapshot of the aircraft impact on the second of the Twin Towers in New York on September 11 2001, whereby some people claim to see the face of the devil in the cloud of smoke it produced;

1 Helmut Federle, *Asian Sign*, 1980,
234 cm x 288 cm, collectie Museum für
Gegewartskunst, Basel

1 Helmut Federle, *Asian Sign*, 1980,
234 cm x 288 cm, Museum für Gegewartskunst
collection, Basel

ANGST ALS ASSOCIATIEVE RUIMTE
Moritz Küng

Zowel angst (gevoel) als ruimte (feit) is een nogal rekbaar begrip. Angst is
een onbehagen, een niet goed definieerbare emotie, die niet voor iedereen
op dezelfde manier of onder dezelfde omstandigheden wordt ervaren.
Ruimte daarentegen is een concreet gegeven en kent uiteenlopende
definities en contexten: ruimte is zowel kosmisch, geometrisch, virtueel,
psychisch, transcendentaal en architectonisch, alsook historisch, sociaal,
ambulant, privé of publiek. De Duitse fenomenoloog en architect Franz
Xaver Baier heeft een hele bundel gewijd aan de definitie van ruimtelijk-
heid, waarin hij in drie hoofdstukken (de niet-geleefde, de geleefde en
de expansieve ruimte) 45 aspecten van ruimtelijke schikkingen beschrijft.[1]
Hieruit blijkt dat gevoel en werkelijkheid zich niet laten scheiden en dat
ze talloze associaties toelaten juist omdat ze op een subjectieve ervaring
berusten.

2 2

Maar in hoeverre is er daadwerkelijk een relatie tussen angst en ruimte,
hoe kunnen ze met elkaar in verband worden gebracht en welke conclusies
kan een architect daaraan verbinden? Want een architect heeft vanuit zijn
discipline de neiging om ruimtes te optimaliseren, en als zich een moment
van angst voordoet, zal hij dit eerder onderdrukken dan expliciet zichtbaar
maken.

Een mogelijke invalshoek is om architectuur per definitie als 'gevaarlijk' of
'beangstigend' te definiëren. Een architect dringt immers met zijn persoonlijke visie op de omgeving of op het toekomstige functioneren daarvan aan
de gemeenschap (gevraagd of ongevraagd) een bepaalde manier van
denken of ideologie op. Architectuur staat letterlijk altijd in de weg en is
per se een obstakel. Architectuur van een groot algemeen belang (een
station, een gerechtsgebouw) zal gemakkelijker worden geaccepteerd dan
architectuur met louter een privébestemming. Naast de functionaliteit is het
ook de esthetische gedaante die de publieke acceptatie al dan niet ver-
hoogt. Architectuur met een groot symboolgehalte kan identiteitversterkend
zijn. Men denkt bijvoorbeeld aan de Eiffeltoren in Parijs, die werd gebouwd
werd naar aanleiding van de honderdste verjaardag van de Franse Revolutie, of recenter aan het Guggenheim Museum in Bilbao, dat een economische impuls gaf aan een hele regio. Hoe groter de *common sense*, de
herkenbaarheid en eigenheid van architectuur, hoe meer de maatschappij
zich hiermee identificeert en zichzelf deze toe-eigent.

Maar architectuur kan naargelang de verandering van de context plots en
ongewild als negatief worden ervaren. Wat ooit werd geprezen als de leef-
wijze van de toekomst – denk aan de Bijlmer in Amsterdam – is door ver-
anderde behoeftes vandaag in een getto veranderd. Of wat ooit stond voor
technologische vooruitgang, welvarende economie en politieke macht
(denk aan de Twin Towers in New York), is vandaag een symbool voor
onmacht en vrees.

Maar waardoor ontstaat het gevoel van angst? Het vreemde en onbekende?
Het niet-controleerbare? Het monumentale? Het tweeduidige? Kan een
beeld, een gebeurtenis, een ruimte, een symbool op zich bepaalde angsten
oproepen? Ik probeer dit te doorgronden met een willekeurige opsomming
of associatieve aaneenschakeling van illustraties.

–   De exorbitante monumentaliteit van de 70 meter hoge pleinsculptuur
    Faro del Comercio, (1984) van Luis Barragán in Monterrey (Nuevo
    Léon) in Mexico, een dubbele, roodbeschilderde betonnen schijf,
    die als abstract-geometrische vorm extreem met het landschap
    contrasteert.

2 New York, 11 September 2001
2 New York, September 11, 2001

- Helmut Federle's painting *Asian Sign* (1980) (Museum für Gegenwartskunst, Basel), which refers to the swastika's original meaning as a symbol of fertility;

- Alberto Giacometti's sculpture *Cube (Pavillon nocturne)* (1934), a black solitaire that exudes a mysterious power;
- A 1995 photo by Aglaia Konrad, which shows an unfinished building in Mexico DF, which diagonal struts saved from collapse after the 1985 earthquake;
- Bruce Nauman's sculpture *Double Steel Cage Piece* (1974) (Museum Boijmans van Beuningen, Rotterdam), which both protects and screens off, seems both transparent and closed;

- Felix Gonzalez-Torres' *Untitled (Fear)* (1991), a mirror in blue glass, in which the viewer sees himself as a picture of fear, an allusion to Michel Foucault's definition of heterotopia (a place where one is, but has no physical access to);
- *l'Espace infinie*, 2002 (destroyed), a monumental sculpture by Ann Veronica Janssens, which literally visualizes an unfathomable depth, a void;
- *Black Suprematist Square* (1914-1915), by Kasimir Malewich (Tretyakov Gallery, Moscow), an icon of radicality in recent painting;

- Dirk Braeckman's photo *G.P.B.X.(2)98* (1998) in which an ordinary conventional-looking door becomes a symbol of hidden madness;
- A detail from the installation *Ur-Haus Rheydt* (1989-1993) by Gregor Schneider, who obsessively reconstructs his own house in different places in new labyrinthine and claustrophobic arrangements;
- A late work by Joseph Beuys, *Hinter dem Knochen wird gezählt, Schmerzraum* (1983) (Fundació 'la Caixa', Barcelona), in which a room is clad entirely in lead and both protects and isolates the viewer from the outside world.

3 3

4 4

Although these examples are not representative, one thing does become clear. The portrayal of fear often appears to be linked to an aesthetic rendering or a pure form. Colour is hardly to be found, and darkness predominates (Braeckman, Giacometti, Konrad, Malewich). Where there is some sign of colour, it evokes associations with alertness (Baragan), suicide (Federle) or schizophrenia (Gonzalez-Torres). The materials are 'modernist' (steel, glass and concrete) and when they are soft (plaster in Janssens' case, lead in Beuys') they have a sense of oppression.
At any rate, what this series of images makes clear to me is that the feeling of fear and space is ultimately linked to the depiction of beauty. The purity, immaculacy or 'naked' and unaffected form enables us to sense an instability which in the end has a confrontational effect and is emotionally disappointing. Good art and architecture does this by definition.

Moritz Küng is a curator. He is responsible for the architecture programme of the arts centre deSingel in Antwerp.

1 Franz Xaver Baier, *Der Raum. Prolegomena zu einer Architektur des gelebten Raumes*, Köln, 1996.

3 Aglaia Konrad, *Mexico City*, 1995, foto
4 Bruce Nauman, *Double Steel Cage Piece*, 1974, collectie Boijmans van Beuningen Museum, Rotterdam

3 Aglaia Konrad, *Mexico City*, 1995, photo
4 Bruce Nauman, *Double Steel Cage Piece*, 1974, Boijmans van Beuningen Museum collection, Rotterdam

- Een momentopname van de vliegtuiginslag op 11 september 2001
  in de tweede Twin Tower in New York, waarvan sommigen beweren
  dat zij in de rookwolk het gelaat van de duivel herkennen.
- Het schilderij *Asian Sign* (1980) van Helmut Federle (Museum für
  Gegenwartskunst, Bazel), dat verwijst naar de oorspronkelijke
  betekenis van de swastika, namelijk als vruchtbaarheidssymbool.

- De zwarte sculptuur *Cube (Pavillon nocturne)* (1934) van Alberto
  Giacometti, die een geheimzinnige kracht uitstraalt.
- Een in 1995 door Aglaia Konrad gemaakte foto, die een onafgewerkt
  gebouw in Mexico toont dat na de aardbeving in 1985 door diagonale
  stutbalken voor instorting werd behoed.
- De sculptuur *Double Steel Cage Piece* (1974) van Bruce Nauman
  (Museum Boijmans Van Beuningen, Rotterdam), die tegelijkertijd
  beschermt en afschermt, transparant en gesloten lijkt.

- *Untitled (Fear)* (1991) van Felix Gonzalez-Torres, een spiegel in blauw
  glas, waarin de beschouwer zichzelf ziet als afbeelding van angst,
  een toespeling op Michel Foucaults definitie van de heterotopie
  (een plek waar men is, maar geen fysieke toegang toe heeft).
- *L'Espace infinie* (2002, vernield), een monumentale sculptuur van
  Ann Veronica Janssens, die letterlijk een peilloze diepte, een niets
  visualiseert.
- *Zwart suprematistisch vierkant* (1914–1915) van Kasimir Malevich
  (Tretjakov Museum, Moskou), een icoon van radicaliteit in de moderne
  schilderkunst.

- Dirk Braeckmans foto *G.P.B.X.(2)98* (1998), waarop een gewoon
  burgerlijk ogende deur een symbool wordt voor verborgen waanzin.
- Een detail uit de installatie *Ur-Haus Rheydt* (1989–1993) van
  Gregor Schneider, die op een obsessieve manier zijn eigenhuis
  reconstrueert op andere plekken in een nieuwe labyrintische en
  claustrofobische schikking.
- Een laat werk van Joseph Beuys, *Hinter dem Knochen* wird gezählt,
  Schmerzraum (1983) (Fundació "la Caixa", Barcelona), een volledig
  met lood beklede ruimte die de beschouwer tegen de buitenwereld
  beschermt, maar tegelijkertijd ook daarvan afzondert.

Hoewel deze voorbeelden niet representatief zijn, wordt toch een gegeven
duidelijk: de voorstelling van angst blijkt vaak aan een esthetische voorstelling of een zuivere vorm gekoppeld te zijn. Kleur is nagenoeg afwezig
en duisternis overheerst (Braeckman, Giacometti, Konrad, Malevich).
Is er sprake van kleur, dan wekt deze associaties met alertheid (Barragán),
suïcide (Federle) of gespletenheid (Gonzalez-Torres). De materialen zijn
'modernistisch' (staal, glas en beton), en als ze zacht zijn (gips bij Janssens,
lood bij Beuys) is hun betekenis beklemmend.
Wat deze reeksen van beelden althans voor mijzelf verduidelijken, is dat
het gevoel van angst en ruimte uiteindelijk gekoppeld gaat met de voorstelling van schoonheid. De puurheid, reinheid of 'naakte' en ongekunstel-
de vorm doet een labiliteit voelen die uiteindelijk confronterend werkt en
emotioneel ontgoochelt. Goede kunst of architectuur doet dit per definitie.

*Moritz Küng is curator. Hij is verantwoordelijk voor de architectuur-
programmatie in kunstencentrum deSingel te Antwerpen.*

1 Franz Xaver Baier, *Der Raum. Prolegomena zu einer Architektur des gelebten Raumes*, Keulen 1996.

5 5

6 6

5 Ann Veronica Janssens, *L'espace infini*, 2002,
540 cm x 420 cm x 310 cm (vernield)
6 Felix Gonzalez-Torres, *Untiteld (Fear)*, 1991,
7,9 cm x 65,7 cm, privé collectie, Italië

5 Ann Veronica Janssens, *L'espace infini*, 2002,
540 cm x 420 cm x 310 cm (destroyed)
6 Felix Gonzalez-Torres, *Untiteld (Fear)*, 1991,
77.9 cm x 65.7 cm, private collection, Italy

Fotomontage 'Fear of the City',
papier op karton
Photomontage 'Angst voor de stad',
papier on carbord

Angst voor de stad   Untitled   93

FEAR OF THE CITY

Rotterdam's public surface reveals a pattern attuned to the comfort of its inhabitants. Ticket machines, cash points, pavement security elements, video cameras, paid parking places, fences, poles and signs, 'neighbourhood watches', rubber-tiled playgrounds … all are the sediment of 'certainty' through which the structures of contemporary society protect themselves. These devices create an illusion of security on an individual level as well. Comfort and fear come together in a sensation of status anxiety; the city dweller's individual universe provides comfort and overcomes fear in a state of permanent ignorance.

In a curious counterpoint to the accumulation of comfort and anxiety on the ground, Rotterdam finds liberation in its phantasmagoric skyline. Spindly and aggressive, the Potemkin skyline unabashedly displays in its appearance its longing to become a city. From a distance, Rotterdam's skyline promises an urban intensity, but once inside the city, at ground level, the intensity is gone. What remains is an urban scenography replete with implements of control.

In essence, the project is positive: to (unabashedly) embrace Rotterdam's dream of creating a city of towers and to redirect this power to reshape the public domain. In this project of 'towers and squares', the main question becomes how the tower volumes relate to the (public) surface level.

Nearly every planned building in Rotterdam is a tower. The tower is the city's main visual value. The idea of the tower as a spatial container has evolved from the recognition of the tower as the a priori type for any development. The tower is the embodiment of the primacy of image – the skin of the building is disconnected from its content to such an extent that it can contain anything, even public space.

ANGST VOOR DE STAD

Rotterdam's publieke maaiveld onthult een structuur die is af-
gestemd op het comfort van haar bewoners. Kaartjesautomaten,
geldautomaten, trottoirbeveiligingselementen, videocamera's,
betaald parkeren, omheiningen, palen en borden, 'buurtwachten',
speelplaatsen met rubbertegels… zijn allemaal het sediment van
'zekerheid' waarmee de structuren van de hedendaagse samenleving zichzelf beschermen. Deze hulpmiddelen creëren ook op
individueel niveau de illusie van veiligheid. Comfort en angst
komen bij elkaar in een gevoel van statusangst; het individuele
universum van de stedeling biedt comfort en overwint de angst
in een staat van permanente ontkenning.

Als merkwaardige keerzijde van de overdaad van comfort en
angst op de grond, vindt Rotterdam zijn bevrijding in zijn fantas-
magorische skyline. De Potemkin-skyline is rank en agressief en
toont in zijn uiterlijk schaamteloos het verlangen een stad te
worden. Vanuit de verte biedt de skyline van Rotterdam de belofte
van stedelijke intensiteit, maar eenmaal in de stad, op de grond,
blijft hier niets van over. Wat resteert is een stedelijke scenografie
vol controlemeubilair.

Het project is in essentie positief: het (schaamteloos) omarmen
van Rotterdam's verlangen een stad van torens te maken en
deze kracht aanwenden om het publieke domein opnieuw vorm
te geven. In dit project van 'torens en pleinen' is de hoofdvraag:
hoe verhouden de torenvolumes zich tot het (publieke) grond-
oppervlak?

Bijna elk gepland gebouw in Rotterdam is een toren. De toren
is haar belangrijkste beeldwaarde. Vanuit de onderkenning van
de toren als a-priori type van elke nieuwe ontwikkeling is de idee
ontwikkeld van de toren als ruimtelijke container. De toren is
de belichaming van het primaat van het beeld; de huid van het
gebouw staat zodanig los van de inhoud dat om het even wat
erin kan worden ondergebracht; zelfs publieke ruimte.

TWO PROJECTS
ON TOWERS AND SQUARES

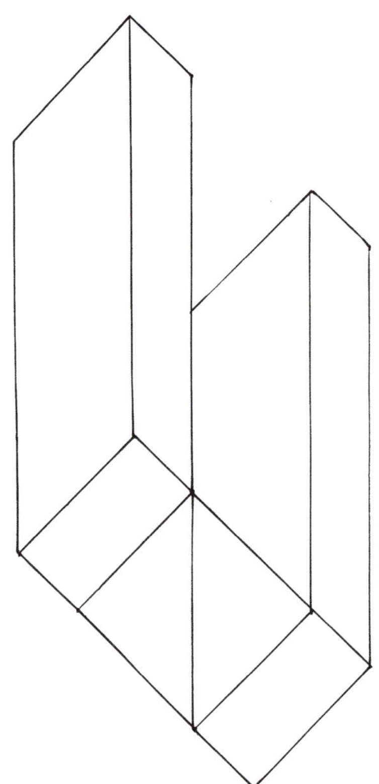

Two Towers and One Square

Two identical towers form a square of 28.8 x 28.8 m.
The towers measure 28.8 x 14.4 on the base and are 72.9 m high.
They are mirroring across the square they form. Each tower is built
on an identical double-cross 'core'. The vertical core incorporates
the tower's structure, installations, elevators and parking space.
Between the reflective façades facing the square the towers'
double cross reveals the pedestrian elevators to the different flats.
The twinned housing towers frame the space in between, the one
their skyline anticipates.

TWEE PROJECTEN
OVER TORENS EN PLEINEN

Twee torens en een plein

Twee identieke torens vormen een plein van 28,8 bij 28,8 meter.
De torens zijn 72,9 meter hoog en meten 28,8 bij 14,4 meter.
Ze zijn gespiegeld over het plein dat ze vormen. Elke toren is
gebouwd op een identieke dubbel-kruiskern. Deze kern incor-
poreert zowel de structuur, de installaties alsook liften en parking.
Tussen de reflecterende façades die het plein vormen, tonen de
kruisvormige schachten de (publieke) liften tot de verschillende
flats. De 'getwinde'/gekloonde woningtorens kaderen die lege
ruimte tussenin die door hun skyline wordt geanticipeerd.

Doorsnede door vloeren en kern
twee torens en een plein
Grote Markt, Rotterdam, 2004

Section through floors and the core
two towers and one1 square
Grote Markt, Rotterdam, 2004

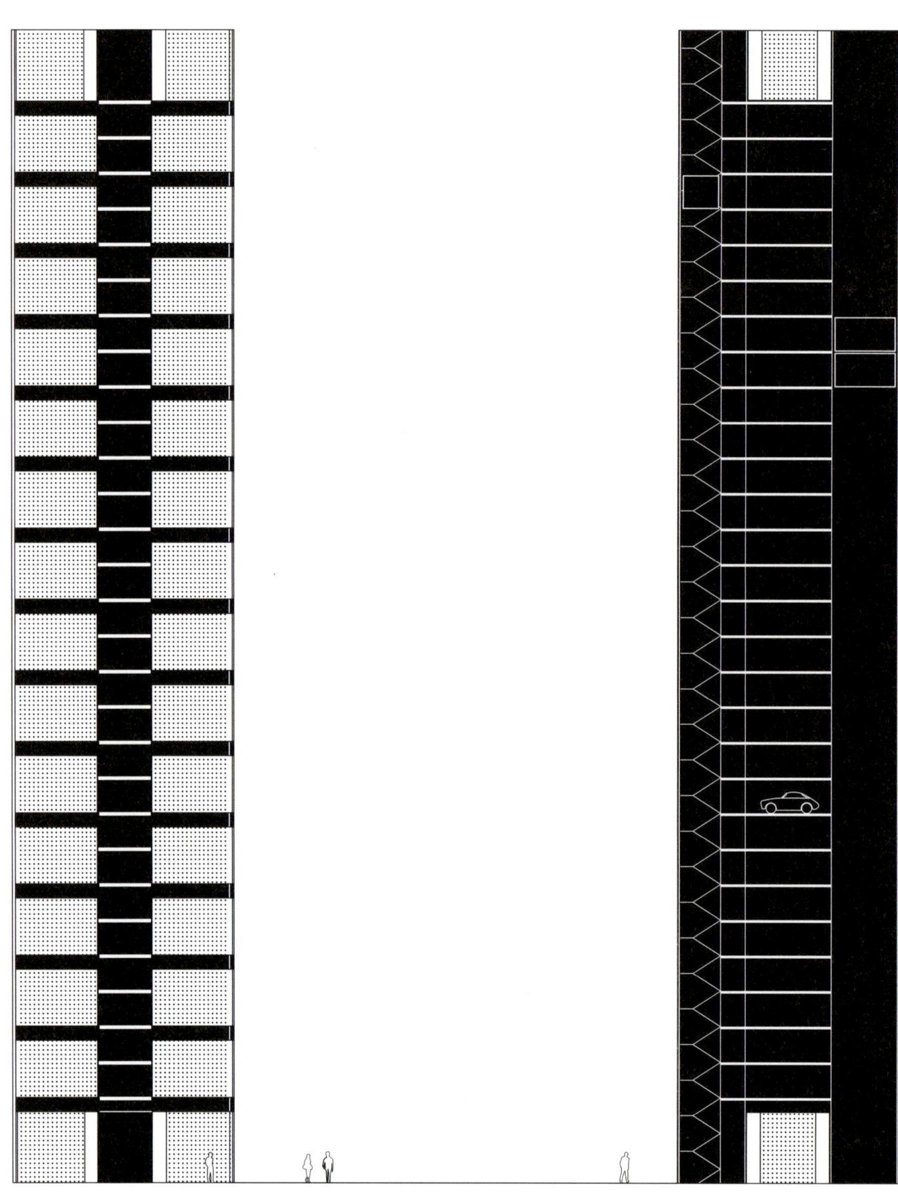

Plattegrond standaardverdieping
twee torens en een plein
Grote Markt, Rotterdam, 2004

Typical floor plan
two towers and one square
Grote Markt, Rotterdam, 2004

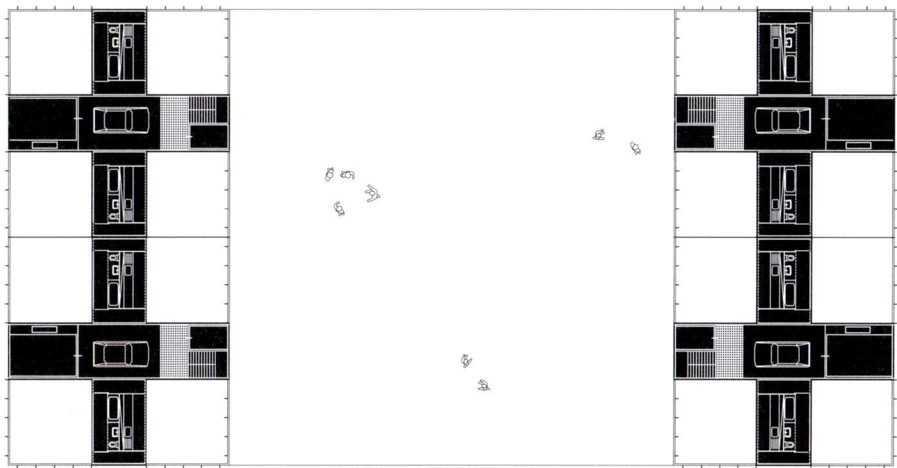

Model (aluminium, mdf, plexiglas),
twee torens en een plein, 2004
Model (aluminum, mdf, plexiglas),
two towers and one square, 2004

Fotomontage (1:1), twee torens en een plein,
Grote Markt, Rotterdam, 2004
Photomontage (1:1), two towers and one square,
Grote Markt, Rotterdam, 2004

TWO PROJECTS
ON TOWERS AND SQUARES

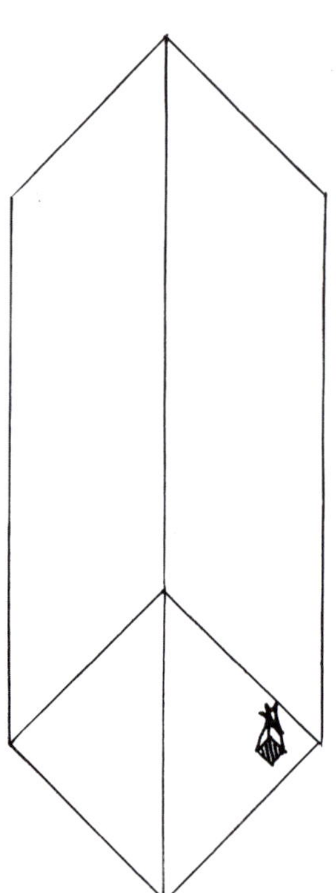

One Tower and One Square

A structure of L-shaped ramps forms a square of 28.8 x 28.8 m.
It is bordering the square and it is 72.9 m high. This structural
perimeter is constructed as a double helix ramp: the inner and
the outer spiral connect seamlessly at the top of the tower.
Both helixes are touching a reflective glass surface in between.
The tower volume is placed over the square with the gigantic
Zadkine sculpture of a man without a heart. The 'Monument for
a Destroyed City', can be seen from afar: from the spiralling
tower above and as a hollowed-out tower in the skyline.

TWEE PROJECTEN
OVER TORENS EN PLEINEN

Een toren en een plein

Een structuur van L-vormige hellingbanen vormt een plein van
28,8 bij 28,8 meter. De rand van het plein is een 72,9 meter hoge
structuur. De structurele omtrek is opgebouwd uit een dubbele
helix helling: de binnen en de buiten spiraal zijn naadloos met
elkaar verbonden aan de top van de toren. Tussen de beide
helixen is een reflecterend glasoppervlak geplaatst. Het holle
torenvolume is geplaatst over het plein met de gigantische
Zadkine sculptuur van een man zonder hart. Het 'Monument
voor een verwoeste stad' is van veraf zichtbaar: van bovenop
in de spiraal toren, en als een uitgeholde toren in de skyline.

Doorsnede
een toren en een plein
Zadkine plein, Rotterdam, 2004

Section
one tower and one square
Zadkine square, Rotterdam, 2004

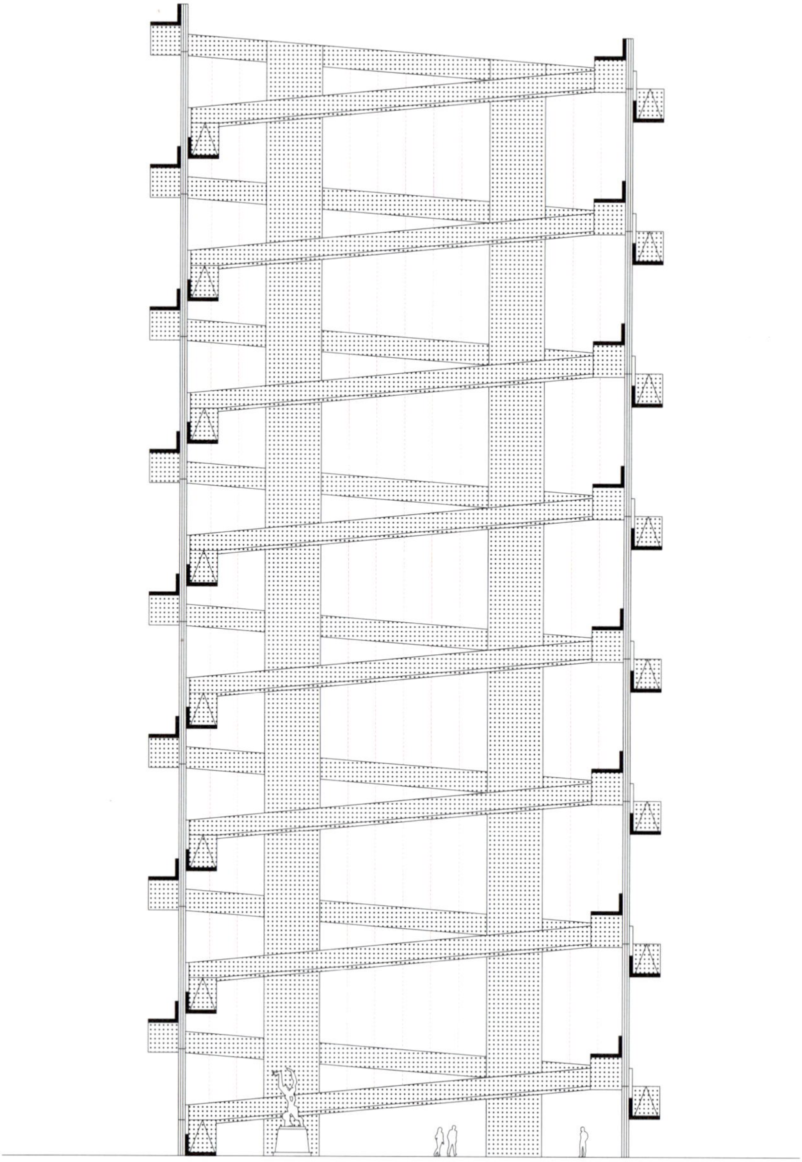

Plan typeverdieping
een toren en een plein
Zadkine plein, Rotterdam, 2004

Floor plan
one tower and one square
Zadkine square, Rotterdam, 2004

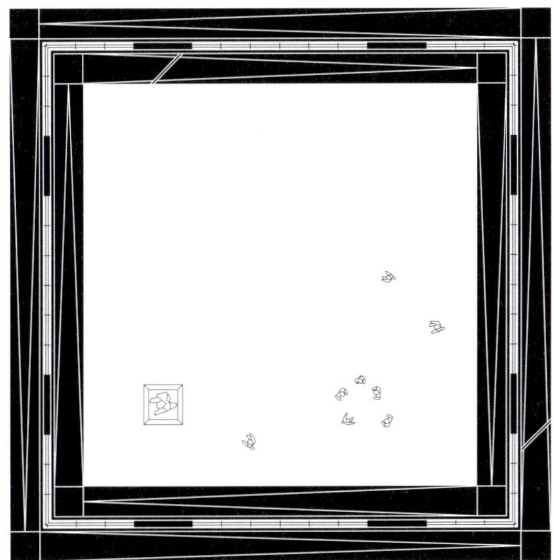

Model (aluminium, plexiglas),
een toren en een plein, 2004
Model (aluminum, plexiglas),
one tower and one square, 2004

Fotomontage (1:1), een toren en een plein,
Zadkine plein, Rotterdam, 2004
Photomontage (1:1), one tower and one square,
Zadkine square, Rotterdam, 2004

A conversation about intentions between Moritz Küng and Untitled:
Kersten Geers, Bas Princen, David Van Severen, Milica Topalovic

FEAR, CITY, LANGUAGE
Architecture and City

MK: Although the theme is fear and space, your work focuses on the idea of fear of the city.

KG: You give the project a certain direction as soon as you manipulate the given theme. In our initial text we made it in two steps from 'fear and space' via 'fear of space' to 'fear of the city'. We no longer view the theme based on the general concepts of 'Fear' and 'Space'; it is not about surveillance cameras and the like.
Our specific subject is an analysis of the increasingly fragmenting city. It is about Rotterdam as an example of a city rebuilt from the ground up after the Second World War.
Rotterdam, with its Potemkin skyline, is built to be visible from as far off as possible. At the same time, it is a city devoid of any specially designed public spaces in the traditional sense of the word.
At the start of this project we wrote 'Where is the public behind the scenes of the city?'
We also wrote that for too long we kept analysing the city by mapping it, looking for subversive structures, characteristic marginal phenomena, just so we could go on believing that there is still a significant public domain behind the scenes. Having desperately tried to rediscover that city, we started having doubts.

MK: Did you rediscover the city?

KG: Well, no…

BP: We did discover parts of it that might in time become a collection of different cities…

MT: …or a layering of different 'cultures'. This is a very specific approach. There was a moment when interest in architecture reached a low point and urban research took off.

MK: And why was that interest so minimal? It is an interesting point.

DvS: Perhaps because there was a fear of space as such…

KG: Indeed, this mapping approach was quite successful, but even though it uncovered certain phenomena or self-regulating structures, it remained an emblematic collection of data that – even when you interpret the word 'city' quite liberally – took on an aesthetic independence and rhetorical objectivity of their own.

MT: 'Generative processing' and 'data-scaping' were attempts to translate this back into architecture. It became an excuse for creating a form.

MK: So, architecture as urban planning?

MT: Exactly. Architecture as the product of a kind of scientific process, as validation…

DvS: Personal decisions were no longer necessary…

KG: It seemed very difficult to turn the 'mapping' approach into a positive design; often it remained very analytical, rather than producing something; it did not lead to proposals – at the cost of one of the most important tools of architects and designers – the chance to intervene in space or in form, to create something and to see what happens.

1 1

2 2

3 3

1   De stad Rotterdam: het
        stadscentrum na de ver-
        woesting van de Tweede
        Wereldoorlog
2,3   Stad van torens: zo goed
        als elk gepland gebouw in
        Rotterdam is een toren

1   City of Rotterdam: the city
        centre after the Second
        World War destruction
2,3   City of towers: nearly
        every planned building in
        Rotterdam is a tower

Een gesprek over intenties tussen Moritz Küng en Untitled: Kersten Geers,
Bas Princen, David Van Severen, Milica Topalovic

ANGST, STAD, TAAL
De architectuur en de stad

4 4

MK: Hoewel het thema angst en ruimte is, gaat jullie werk over het onderwerp van angst voor de stad.
KG: Zodra je het gegeven thema manipuleert, geef je het project een
bepaalde richting. In onze begintekst kwamen we in twee stappen van
'angst en ruimte', via 'angst voor ruimte' bij 'angst voor de stad'. We
beschouwen het thema niet langer vanuit de algemene begrippen 'angst'
en 'ruimte'; het gaat niet over zaken als beveiligingscamera's.
Ons specifieke onderwerp is een analyse van de steeds verder versplinterende stad. Het gaat over Rotterdam als voorbeeld van een stad die na de
Tweede Wereldoorlog van de grond af is herbouwd.
Rotterdam, met zijn Potemkin-skyline, is gebouwd om van zo veraf mogelijk zichtbaar te zijn. Tegelijkertijd is het een stad die is gespeend van elke
vorm van speciaal ontworpen openbare ruimte in de traditionele betekenis
van het woord. Aan het begin van dit project schreven we: 'Waar is het
publiek achter de coulissen van de stad?'
We schreven ook dat we de stad te lang zijn blijven analyseren door haar
in kaart te brengen, door te zoeken naar subversieve structuren, typerende
marginale verschijnselen, om maar te kunnen blijven geloven dat er nog
altijd een publiek domein van betekenis schuilgaat achter die coulissen.
Nadat we wanhopig hadden gepoogd die stad te herontdekken, begonnen
we te twijfelen.
MK: Heb je de stad herontdekt?
KG: Nou, nee…
BP: We hebben delen ervan herontdekt die mettertijd wellicht een
verzameling van verschillende steden zouden kunnen vormen…
MT: … of een gelaagdheid van verschillende 'culturen'. Dit is een heel
specifieke aanpak. Er is een moment geweest dat de belangstelling voor
architectuur tot een minimum was gedaald en het stedelijk onderzoek
explosief toenam.
MK: En waarom was die belangstelling zo gering? Dat is een interessant
punt.
DvS: Misschien omdat er sprake was van angst voor ruimte als zodanig…
KG: Zeker, die aanpak van in kaart brengen had veel succes, maar hoewel
daarmee bepaalde verschijnselen of zelfregulerende structuren zijn ontdekt,
bleef het bij een emblematische verzameling gegevens die – zelfs als je de
betekenis van het woord 'stad' ruim interpreteert – een eigen esthetische
zelfstandigheid en retorische objectiviteit gingen aannemen…
MT: Met 'generative processing' en 'data-scaping' is een poging gedaan
dit terug te vertalen in architectuur. Het werd hier een excuus voor de
totstandbrenging van de vorm.
MK: Architectuur als stedenbouw dus?
MT: Precies. Architectuur als product van een soort wetenschappelijk
proces, als validering…
DvS: Persoonlijke beslissingen waren niet langer noodzakelijk…
KG: Het leek heel moeilijk de benadering van 'in kaart brengen' om te
zetten in een positief ontwerp; het bleef vaak erg in het analytische steken
in plaats van dat het iets opleverde, het leidde niet tot voorstellen. Dit ging
ten koste van een van de belangrijkste gereedschappen van architecten en
ontwerpers: de mogelijkheid te interveniëren in ruimte of vorm, iets te
maken en te kijken wat er gebeurt.

4   Stad van torens:
Rotterdams verlangen naar
een skyline omarmen

4   City of towers: embracing
Rotterdam's desire for a
skyline

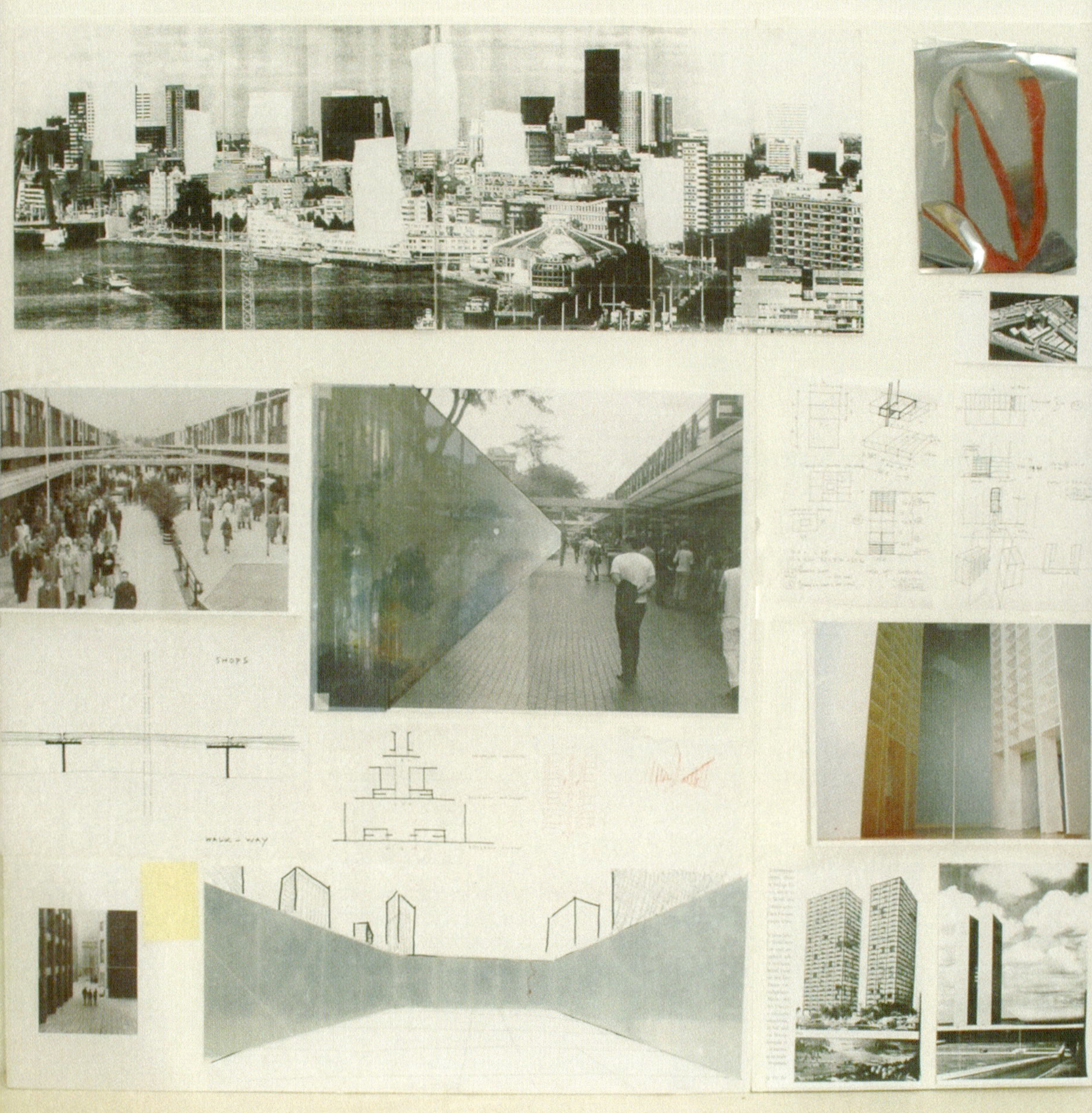

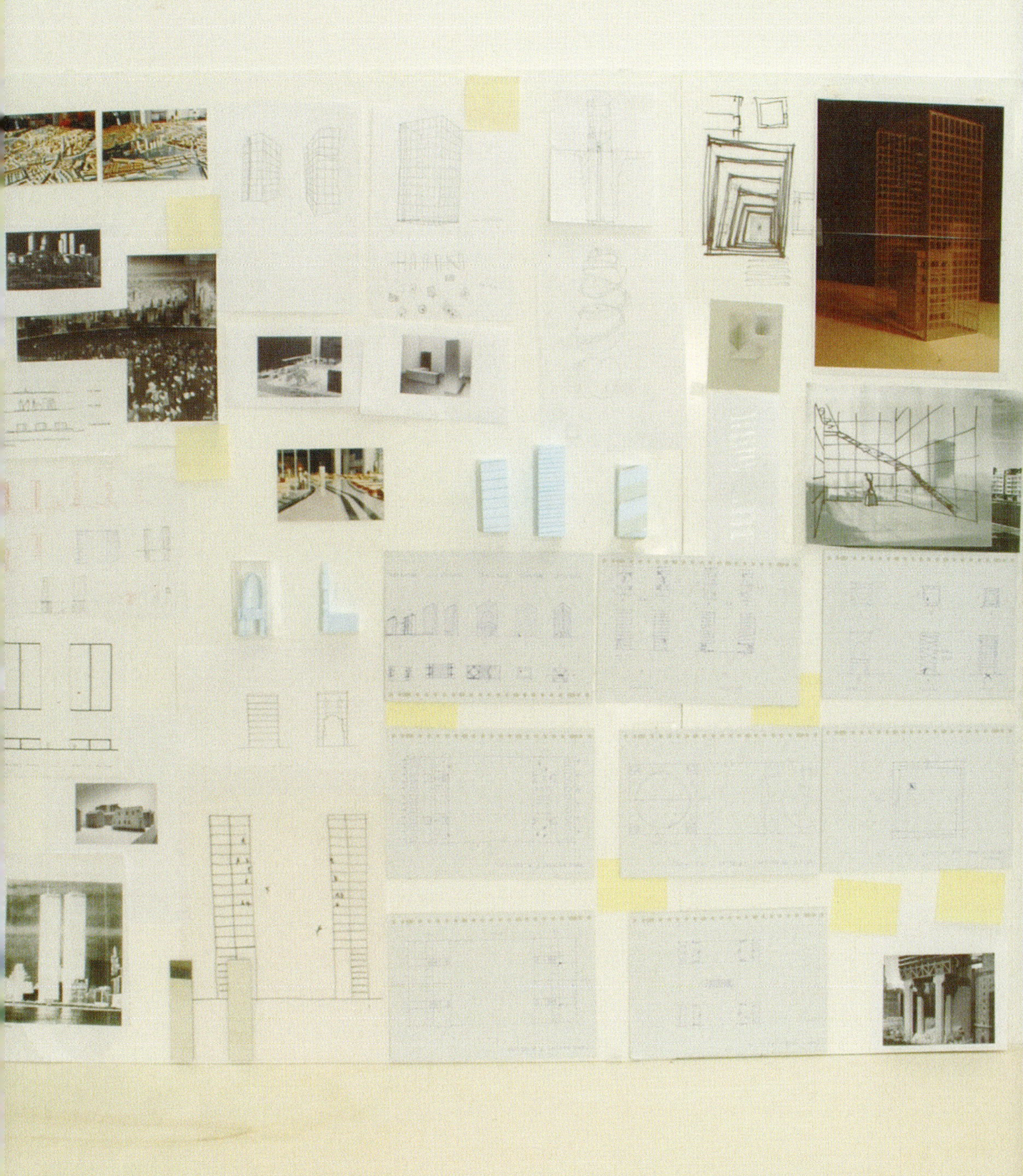

Artist Architect

KG: We wanted to place more emphasis on the project as a method of research once again; we wanted to view a project, whether an actual project or one on paper, as a way to conduct research in any case. We studied examples, with a certain naïveté; an intervention, in itself, seems a good way to tackle larger urban issues. Some of Dan Graham's projects address issues of suburbanization in a powerful way. Later our interest was drawn by how Daniel Buren approached certain projects and how these were received by critics. In his text 'In-situ versus the ready-made' Guy Lelong explained Buren's in-situ as an intervention, a thing, an object, the opposite of the ready-made, which remains only a concept, a verbal formulation that can be reduced to the word itself.
Gradually we shifted our attention to the visual arts: the works of minimalist artists (Lewitt), post-minimalist conceptual artists (Graham) and interventionist artists, and as we talked about last time, the newer generation of artists who create personal interventions which are at the same time very spatial, such as Dominique-Gonzales Foerster or Thomas Demand. Of course, we are architects, not artists. We tried to keep in mind that we work on a very different scale. Yet we still didn't know how to regain the confidence to deal with space and the city.
We try not to wallow in melancholy, but on the other hand we also try to avoid the precisely opposite approach, which you could call 'programming'. Space is now often produced only purely for its commercial value, focusing on the quantity of square meters of façade, etcetera. The question for us became whether we could we do projects with a certain architectural autonomy?
MK: To me, the major difference between an architect and an artist is that an architect works within a heteronomous field, not an autonomous one. In my opinion, an architect constantly tries to construct frames of reference within which he can operate and from which his work results. In this way an artist is very different from an architect.
What you are criticizing, on the other hand, seems to be a logical aspect of the position of an architect. By this I mean that an architect appropriates theories from others, or another identity – that's just a question of strategy.
MT: You might say this viewpoint, whereby strategy becomes more important than the work itself, has gained a lot of ground in the last 10 years. The paradox is that the strategy manifests itself as architecture and architecture gets lost within it.

Towers and Squares

KG: Gradually we developed an interest in towers and squares as archetypes of architectonic language. They stand for right and left, maximum profit and minimum profit, private and public, vertical and horizontal, volume and expanse, density and emptiness. By working with these two forms, we hope to broach certain themes that relate to the construction of a city. Let's put it this way: if you make an 'empty' square, does that make it more public or less public?
MK: What I find interesting in the idea of a tower and a square is that a tower has always been a symbol of economic activity and corporate capitalism, but since 9/11 a tower is also a symbol of fear. Similarly, a square is a symbol of collectivity and therefore it is also a symbol of fear.
MT: We were talking earlier about how there is a principle today that says public space should be either commercial or green. It is almost impossible to design anything else, let alone defend it. For instance, we make a

5,5

6,6

7,7

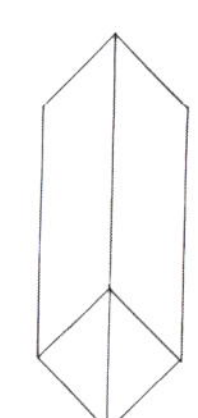

8,8

5,6   Toren en plein, in hun pure incarnatie: vertikaal en horizontaal, hoog en laag, winstgevend en verlieslatend, privaat en publiek, volte en holte
7   Snede en spiegel: torens zijn gespiegeld over het plein dat door hen wordt gevormd
8   Container: een uitgeholde toren kan zelfs publieke ruimte bevatten

5,6   Tower and Square, in their pure idea: vertical and horizontal, high and low, the profitable and the unfeasible, private and public, volume and void
7   Slice and mirror: towers are mirroring each other across the square they form
8   Container: A hollowed-out tower can even contain public space

Kunstenaar Architect

KG: We wilden weer meer nadruk leggen op het project als onderzoeks-
methode, we wilden een project – op papier of concreet – beschouwen als
een manier om hoe dan ook onderzoek te doen.
We hebben vanuit een zekere naïviteit voorbeelden bestudeerd; een interventie lijkt op zich al een goede manier om grotere stedenbouwkundige
problemen aan te pakken. In bepaalde projecten van Dan Graham worden
onderwerpen op het gebied van suburbanisatie op krachtige wijze aan de
orde gesteld. Later raakten we geïnteresseerd in de manier waarop Daniel
Buren bepaalde projecten aanpakte en hoe deze door de critici werden
ontvangen. In zijn tekst 'In-situ versus the ready-made' interpreteerde Guy
Lelong het in-situ van Buren als een interventie, een ding, een object, de
tegenhanger van de readymade die slechts een concept blijft, een formule-
ring in woorden die kan worden teruggebracht tot niets dan het woord zelf.
Geleidelijk verschoof de aandacht naar de beeldende kunst: de werken van
minimalistische kunstenaars (Lewitt), postminimalistische conceptuele kunstenaars (Graham) en interventiekunstenaars, en, waar we de vorige keer
over spraken, de jongere generatie kunstenaars die persoonlijke interventies maken die tegelijkertijd zeer ruimtelijk zijn, zoals Dominique-Gonzales
Foerster of Thomas Demand.
We zijn natuurlijk architecten, geen kunstenaars. We hebben geprobeerd
er rekening mee te houden dat wij op een heel andere schaal werken.
Maar we wisten nog altijd niet hoe we het zelfvertrouwen moesten herwinnen
om met ruimte en de stad om te gaan.
We proberen niet weg te zinken in melancholie, maar anderzijds proberen
we ook de exact tegenovergestelde aanpak, die je 'programmering' zou
kunnen noemen, te vermijden. Ruimte wordt tegenwoordig vaak louter ge-
produceerd om zijn commerciële waarde, waarbij het gaat om hoeveelheid
vierkante meters gevel en dergelijke.
De vraag werd voor ons of we projecten konden doen die over een zekere
architectonische autonomie beschikten.

MK: Het grote verschil tussen een architect en een kunstenaar is wat mij
betreft dat een architect niet werkzaam is binnen een autonoom maar
binnen een heteronoom gebied. Volgens mij probeert een architect voortdurend referentiekaders te vormen waarbinnen hij kan opereren en waaruit zijn werk voorkomt. Daarin verschilt een kunstenaar enorm van een
architect.
Wat jullie bekritiseren lijkt me daarentegen een logisch aspect van de
positie van een architect. Ik bedoel dat een architect zich theorieën van
anderen toe-eigent, of een andere identiteit; dat is gewoon een kwestie
van strategie.

MT: Je zou kunnen stellen dat die houding, waarbij de strategie belang-
rijker wordt dan het werk zelf, de laatste tien jaar veel terrein heeft
gewonnen. Het paradoxale is dat de strategie zich manifesteert als
architectuur en dat de architectuur erin verloren raakt.

Torens en pleinen

KG: Gaandeweg kregen we belangstelling voor torens en pleinen als de
archetypen van de architectonische taal. Ze staan voor rechts en links,
maximale winst en minimale winst, privé en openbaar, verticaal en horizontaal, volume en uitgestrektheid, dichtheid en leegte. Door met deze twee
vormen te werken, hopen we bepaalde thema's aan de orde te stellen die
betrekking hebben op de bouw van een stad. Laat ik het zo zeggen: als je
een 'leeg' plein maakt, is dat dan meer of minder openbaar?

9 9

10 10

9   Poging tot stedelijk project:
het localiseren van toren en
plein projecten in Rotterdam
10 Poging tot typologisch
project: variaties op ruimte
en programma

9   Urban project attempt:
locating tower and square
projects in Rotterdam
10 Typological project attempt:
variations according to
space and programme

square between two towers and we make a square in another tower which is hollow, and at best a look-out tower – today this sounds almost pathetic; it seems naïve. However, I believe projects are very provocative once they exude the ambience of certain spaces from the past, when the public space was much more seen as an ideal.

In a recent lecture Dan Graham criticized the Rietveld Schroeder House as 'too programmed'. Too much or too little programme is an interesting idea, because it implies thinking in aesthetic terms. In that sense, I wouldn't call our approach purely formal; there is a certain measure with which you interpret the programme, and that leaves room to manoeuvre, as well.

KG: I think architecture is ultimately about space. It has to do with the fact that certain architecture turns programme into rhetoric. We design spaces which are first and foremost spaces, but they house a programme. By keeping them 'open' we hope to provide a response to the predominant production of experience.

MT: The idea of the backstage of the city is derived from a public theatre. We are referring to strictly programmed and supervised public gatherings, like the events that regularly take place throughout Rotterdam.

KG: You could say that an event is a public theatre, the backstage of which is formed by the rest of the city. When we say we want to find the spaces behind the scenes, we mean that we want to create spaces that do not slavishly follow these dynamics.

MT: Like the Rotterdam skyline, the more prominent locations in the city are being turned into an urban scenography, usually focused on market driven identity and image.

(…)

At the start of this project we attempted to create an urban project through architecture. We proposed various tower projects for several building sites in Rotterdam. We then realized that all of the projects were based on two ideas, two fairly abstract types: a container tower with a square inside it and two (mirroring) towers with a square in between.

KG: It's interesting that they were not developed from their specific sites, yet both are quintessentially urban typologies in the way in which they were composed

Although neither project was developed specifically for its site, each became a project of urban design. Ultimately, the tower and square projects are attempts to generate a very precise provocation. They are not contextual, but they engage in a dialogue with the broader context of the city in which they are located.

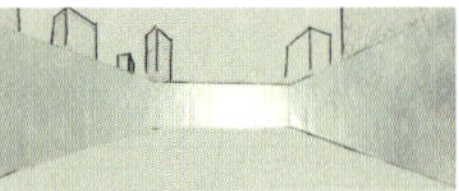

11 11

12 12

13 13

14 14

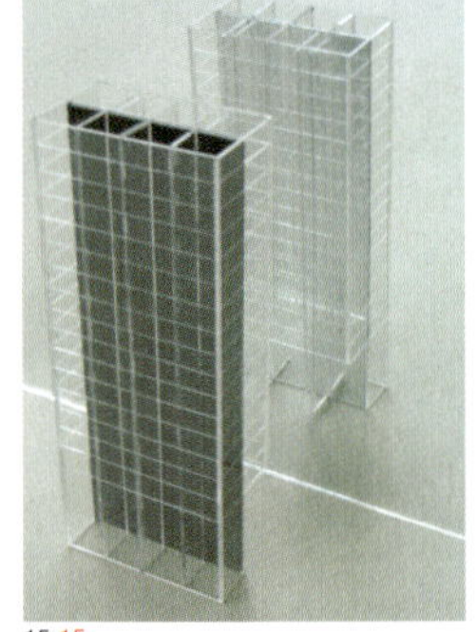

15 15

11   Een ingreep: een plein in een plein
12, 13   Een ingreep: de Lijnbaan, de openlucht schopping mall uit de jaren 50 is een symmetrische gebouwen strip. De winkelstrip wordt doorgeknipt door een smalle spiegelende strook ruimte
14, 15   Suburbaan toren project: gespiegelde toren volumes maken de ruimte ertussenin

11   An intervention: A square in a square
12, 13   An intervention: the Lijnbaan, an open air shopping mall from the 1950s is a symmetrical strip of buildings. The intervention is to cut it through with a slender mirroring strip of space
14, 15   'Suburban' tower project: mirroring tower volumes make a space in-between

MK: Wat ik interessant vind aan de idee van de toren en het plein, is dat de toren altijd een symbool is geweest van economie en bedrijfskapitalisme; maar sinds 9/11 staat de toren ook symbool voor angst. Zo is het plein een symbool van gemeenschappelijkheid en daarmee ook een symbool van angst.

MT: We hadden het er eerder al over dat tegenwoordig het principe bestaat dat openbare ruimte óf commercieel óf groen moet zijn. Het is bijna onmogelijk iets anders te ontwerpen, laat staan dat te verdedigen. We maken bijvoorbeeld een plein tussen twee torens, en we maken een plein in open carrévorm in een andere toren die hol is en in het beste geval een uitkijktoren. Dit klinkt in deze tijd bijna pathetisch, het lijkt naïef. Ik denk echter dat projecten zeer provocerend zijn zodra ze de sfeer ademen van bepaalde ruimtes uit het verleden, toen openbare ruimte nog veel meer als ideaal werd gezien.

Onlangs zei Dan Graham in een lezing over het Rietveld-Schröderhuis dat het 'te geprogrammeerd' was. Te veel of te weinig programmering is een interessant begrip, omdat het impliceert dat er in esthetische termen over wordt nagedacht. In dat opzicht zou ik onze aanpak niet puur formeel willen noemen; er is een zekere maatgeving van waaruit je de programmering interpreteert, en dat laat ook enige speelruimte.

KG: Ik denk dat architectuur uiteindelijk over ruimte gaat. Het heeft te maken met het feit dat bepaalde architectuur programmering omzet in retoriek. Wij ontwerpen ruimtes die allereerst ruimtes zijn, maar er schuilt wel een programma achter. Door ze 'open' te houden, hopen we tegenspel te bieden aan de overheersende productie van ervaring.

MT: Het idee van de coulissen van de stad is ontleend aan een openbaar theater. We doelen daarmee op strikt geprogrammeerde en gecontroleerde openbare bijeenkomsten, zoals de regelmatig terugkerende evenementen overal in Rotterdam.

KG: Je zou kunnen zeggen dat een evenement een openbaar theater is waarbij de coulissen worden gevormd door de rest van de stad. Als we zeggen dat we de ruimtes willen vinden achter die coulissen, betekent dit dat we ruimtes willen maken die deze dynamiek slaafs volgen.

MT: Net als de skyline van Rotterdam worden de meer prominente plaatsen in de stad omgebouwd tot een stedelijke scenografie waarbij het veelal gaat om marktgerichte identiteit en beeld.

(…)

Bij het begin van het project probeerden we met behulp van architectuur een stedelijke project te maken. We deden voorstellen voor diverse torenprojecten op een aantal bouwlocaties in Rotterdam. Toen beseften we dat alle projecten gebaseerd waren op twee ideeën, twee redelijk abstracte typen: een rechthoekige toren met in de toren een plein en twee (spiegelende) torens met een plein ertussen.

KG: Het is interessant dat ze niet zijn ontwikkeld vanuit de specifieke locatie, maar desondanks zijn het alletwee bij uitstek stedelijke typologieën in de manier waarop ze zijn samengesteld.

Hoewel beide projecten dus niet specifiek voor hun locaties werden ontwikkeld, werden het urban designprojecten. Het toren- en pleinproject probeert uiteindelijk om een heel precieze provocatie uit te lokken. Ze zijn niet contextueel, maar gaan een dialoog aan met de bredere context van de stad waarin ze zich bevinden.

16 16

17 17

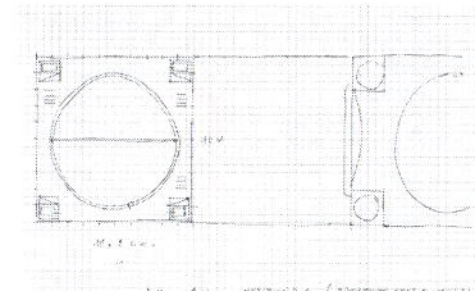

18 18

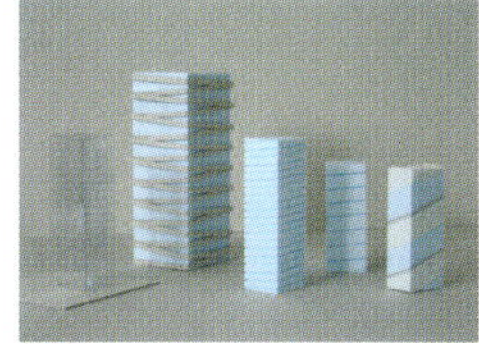

19 19

20 20

16, 17, 18 Publiek toren project: het nieuw torenvolume spiegelt de minaret-torens van de moskee

19 Container toren project: de toren wordt over het plein geschoven, uitgehold. De rand wordt een envelope van ruimte; het maakt de toren het plein

20 Een toren en een plein: Zadkine plein toren

16, 17, 18 Public bath tower project: the new tower volume mirrors the minaret-towers of the mosque

19 Container tower project: moving tower over the square, hollowing it out, making it's border a wrap of space; making the tower the square

20 One tower and one square: Zadkine square tower

TWEAK

With the statement 'fear is illusion',
Shine 5.0 conducts research into the
possibilities of addressing the stereotypes of fear and space in our society
within their design: the dark alley,
the unknown stranger, the emptiness...
Shine 5.0 seeks to connect with myths
about fear, space and themed environments.
Shine 5.0 opted for a specific inner-city
environment and began by conducting
field research there. The research was
carried out in two domains: on the one
hand a spatial analysis of the environment, on the other hand a behavioural
analysis of its users. To Shine 5.0, both
domains are inextricably linked.
A reciprocal relationship exists between
the space and its users: fear is as much
a product of the environment, of stereo-
typing, as it is a product of human
beings, of their rhetoric, their isolation,
their anticipation. The synthesis of
the two angles in the field research
produced elements that could bring
about a different experience of the
environment through reprogramming.
Shine 5.0 rewrites the existing environment through the introduction of a new
scenario, rhetoric, or reading. This does
not relegate the existing context to the
background but rather makes it part
of the new situation. That scenario
provides, temporarily and spontan-
eously, information about the relationship between man and fear; in this
sense fear becomes a representation.
This information or experience can be
further put to use, until a situation is
stripped of the complex patterns of fear
and rhetoric that operate within the
environment. Shine 5.0, by means of
interventions, clarifies a situation that
may or may not be illusory. The interventions make possible the perception
and experience of a situation in which
fear has not vanished but has acquired
clear contours.

Met het statement 'angst is inbeelding'
doet Shine 5.0 onderzoek naar de
mogelijkheden om ontwerpend in te
gaan op de stereotypen van angst en
ruimte in onze samenleving: het donkere
straatje, de onbekende vreemdeling,
de leegte… Shine 5.0 zoekt hier aansluiting bij de mythes over angst,
ruimte en gethematiseerde omgevingen.
Shine 5.0 koos voor een specifieke
binnenstedelijke omgeving en deed daar
in eerste instantie aan veldonderzoek.
Het onderzoek bewoog zich op twee
velden: enerzijds een ruimtelijke analyse
van de omgeving, anderzijds een
gedragsanalyse van de gebruikers
ervan. De beide velden zijn voor Shine
5.0 onlosmakelijk met elkaar verbonden.
Tussen de ruimte en de gebruikers
ervan bestaat een wederzijdse relatie:
angst is zowel een product van de
omgeving, van de stereotypering,
als een product van de mens, van zijn
retoriek, zijn isolatie, zijn anticipatie.
De synthese van beide invalshoeken
in het veldonderzoek leverde elementen
op die door herprogrammatie een
andere beleving van de omgeving
konden bewerkstelligen. Shine 5.0
herschrijft de bestaande omgeving door
de introductie van een nieuw scenario,
retoriek of lezing. Daarbij verdwijnt
de bestaande context niet naar de
achtergrond, maar wordt deze onderdeel van de nieuwe situatie. Dat scenario verschaft, tijdelijk en spontaan,
informatie over de relatie tussen mens
en angst; in die zin wordt angst een
verbeelding. Met die informatie of
ervaring kan worden verdergewerkt,
tot een situatie is ontdaan van de
complexe angstpatronen en retoriek
die door de omgeving heen bewegen.
Shine 5.0 verduidelijkt door middel van
ingrepen een al dan niet ingebeelde
situatie. De ingrepen maken de waarneming en beleving mogelijk van een
situatie waarin angst niet verdwijnt,
maar duidelijke contouren krijgt.

TWEAK
TWEAK
PROJECT:
SHINE 5.0
BIJDRAGE / CONTRIBUTION:
JACOB VOORTHUIS

The Enlightened Wilderness: Tweaking Urban Space
Jacob Voorthuis

A tweak entails a far from straightforward complement of activities and gestures. The OED gives a relatively straight forward definition: 'to seize and pull sharply with a twisting movement.' It backs this definition up with a telling example: 'To pull by the nose as a mark of contempt.' That is where the word's innocence is lost, in the possibility of contempt. In my own mind tweaking conjures up the image of an experienced hostess surveying her living room with a penetrating, military gaze while brushing through and giving a last tug at the curtains and the expensive lush sofa before welcoming her visitors. A third variation, and one I suspect to be rather closer to the purposes of the Shine group, involves the small but confident adjustments to the impressive constellation of knobs on a mixing panel in a studio before the recording is finalized. All three variations will play an implicit and silent role in the following essay which tries to explore the theoretical environment in which the Shine group have launched their TWEAK project.
The essay sets out to do two things. In the first place it is an attempt to undermine the concept of fear as a seemingly well-defined and apparently self-evident praxis. In the second I want to look at the idea of domestication of space. Both of these are necessary, I believe, to make sense not just of the intentions of the Shine group, but the possibility of a wider reading of the project.
Shine's urban intervention attempts, in my view, to do two interesting things. On the one hand it is an exercise in recalibrating the conventional responsibilities of architectural and urban design. In this way they have taken Cedric Price's short sharp thrust to heart: 'It is vital to see where architecture isn't needed.' Their attempt to design a soft architecture, rather than a hard architecture, questions the role of the urban designer, the architect, whereby the required product is not a building, not an urban assembly, but merely a questioning of assumptions, a restructuring of the social space through interventions which have as their sole purpose to make people conscious of themselves and their ideas about things. In this sense they are entering the realm of art. In this project they have chosen to graft the comfortably familiar, an open wall-less living room, into the all too familiar space of modern alienation. In this way their simulated living room, is at one and the same time an existential act of the absurd as well as a gesture with a clean political purpose: (re)domesticating the modern sublime. The incongruence of the graft will, they hope, stimulate a re-evaluation of urban spaces as well as our knee-jerk reaction to them.
The idea of the street as an urban living room is of course not new. Walter Benjamin had explored the idea of public space as an interior space, the street as a living room. Louis Kahn described the street as a room by agreement. Aldo van Eyck wanted the city to be a house and vice versa. People in rather more kindly climates than the Dutch have been living the idea of a street as a living room since time immemorial, since the first street was conceived perhaps in Khirokitia, Cyprus. More recently Koolhaas placed a wall-less living room in the large empty agora of his conference centre. Mies van der Rohe, Philip Johnson and Shigeru Ban have explored wall-lessness in their houses, inviting a similar confrontation with the outside. The designers of modern nomadism have often played with the weird intimacy of domestic bliss in unheimisch surroundings; Archigram and Superstudio are potent examples from the 1960s. So the answer might not be new as such, but what is new is the question as well as the context in which the answer is given. And that gives us the peculiar flavour to this project. This open, wall-less urban living room was specifically conjured up to deal with spaces that have become infected by fear.

De verlichte wildernis: het 'fijn afstemmen' van de stedelijke ruimte
Jacob Voorthuis

Met het Engelse woord 'tweaking' wordt een verre van simpel geheel aan
handelingen en gebaren aangeduid. De *Oxford English Dictionary* geeft een
betrekkelijk simpele definitie: 'vastpakken en abrupt trekken met een draaiende
beweging'. Deze definitie wordt geïllustreerd met een treffend voorbeeld:
'aan de neus trekken ten teken van minachting'. Daar verliest het woord zijn
onschuld: in de mogelijkheid van minachting. Bij mij roept 'tweaking' het beeld
op van een ervaren gastvrouw die met een doordringende, militaire blik haar
woonkamer inspecteert, nog even afstoft en de gordijnen en de dure, luxueuze
sofa rechttrekt voordat ze gaat opendoen voor haar visite. Een derde betekenis, die naar ik aanneem dichter bij de doelstellingen van de Shine-groep staat,
betreft de kleine maar trefzekere bijstellingen aan de indrukwekkende knop-
penwinkel van een mengpaneel in een studio, bij de afronding van een plaat-
opname. Alledrie de betekenissen spelen een impliciete en stilzwijgende rol
in dit artikel, dat een verkenning is van de theoretische omgeving waarin de
Shine-groep zijn TWEAK-project heeft gelanceerd.
Dit artikel heeft een tweeledige doelstelling. In de eerste plaats is het een
poging het begrip 'angst' als ogenschijnlijk welomschreven en vanzelfspreken-
de praxis te ondermijnen. Ten tweede wil ik kijken naar de idee van domesticering van de ruimte. Beide zijn volgens mij noodzakelijk, niet alleen om de
doelstellingen van de Shine-groep te kunnen begrijpen, maar ook om hun
project in bredere zin te kunnen interpreteren.
Volgens mij stelt Shine 5.0 zich met zijn stedelijke interventie twee interessante
dingen ten doel. Enerzijds wil de groep de traditionele verantwoordelijkheden
van het architectonisch en stedenbouwkundig ontwerpen herijken. Wat dat
betreft heeft Shine 5.0 de scherpe opmerking van Cedric Price ter harte
genomen: 'Het is van essentieel belang te weten waar géén architectuur
nodig is.' In hun poging een zachte in plaats van een harde architectuur te
ontwerpen, stellen de Shine-leden de rol van de stedenbouwkundig ontwerper
en de architect ter discussie. Daarbij streven ze niet naar een eindproduct in de
vorm een gebouw of een stedelijk ensemble, maar zetten slechts vraagtekens
bij vooronderstellingen en reorganiseren de maatschappelijke ruimte via
interventies die enkel ten doel hebben de mensen bewust te maken van
zichzelf en hoe ze over dingen denken. Hier betreden ze het terrein van de
kunst. In dit project hebben ze ervoor gekozen een prettig vertrouwde, zij het
muurloze woonkamer over te planten naar de maar al te vertrouwde ruimte
van de moderne vervreemding. Zo is hun nagebouwde woonkamer zowel een
existentiële, absurdistische daad als een gebaar met een duidelijk politiek
doel: het (her)domesticeren van de moderne grootsheid. Ze hopen dat de
ongerijmdheid van deze transplantatie ons aanzet tot herbezinning op de
stedelijke ruimtes en onze voorspelbare reacties daarop.
Het concept van de straat als een stedelijke woonkamer is natuurlijk niet nieuw.
Walter Benjamin heeft al de idee verkend van de publieke ruimte als binnen-
ruimte en de straat als woonkamer. Louis Kahn heeft de straat omschreven als
een 'kamer bij consensus'. Aldo van Eyck wilde dat de stad een huis was en
omgekeerd. Mensen die in een wat milder klimaat leven dan het Nederlandse
brengen het concept van de straat als woonkamer al sinds onheuglijke tijden
in de praktijk, sinds de aanleg van de eerste straat, wellicht in Khirokitia op
Cyprus. In een recenter verleden plaatste Koolhaas een muurloze woonkamer
in de grote lege agora van zijn conferentiecentrum. Mies van der Rohe,
Philip Johnson en Shigeru Ban hebben de muurloosheid in hun huizen
onderzocht en daarmee uitgenodigd tot eenzelfde confrontatie met de
buitenwereld. De ontwerpers van een modern nomadenbestaan hebben vaak
gespeeld met de eigenaardige intimiteit van huiselijk geluk in een unheimliche
omgeving; Archigram en Superstudio zijn daarvan sterke voorbeelden uit de

The world has changed. Not just the former colonies, but now Europe itself
is being increasingly confronted with the consequences of the inherent
contradictions of the cultural interface of mere ethnic and religious tolerance
as opposed to positive curiosity and enjoyment. The problems of an often
unilateral multiculturalism, whereby one person tolerates the other, while the
other fails to tolerate the one are being acted out on every doorstep. As one
of the many products of this internal failure we have generated the greatest
media star of all-time: the selfdestructing terrorist, whose altruistic acts of
destruction and murder in the name of his miserable cause has destroyed
the last vestiges of that social simulacrum, the moral border. At the same
time we are being rudely confronted with the paradox of our super-civilized
society. As a result of our enlightened sensibilities, moulded not in the last
place by the pedagogy of Jean Jacque Rousseau we have tried to create
noble savages as children: independent young adults, confident, full of
curiosity and initiative, able to teach themselves naturally from nature and the
natural. We achieved a considerable success. The result, however, is made
complicated by the large number of rather more savagely inclined nobles,
flooded with confusing hormones and revisiting the macho as a way to shape
their image and satisfy their craving for respect. From the same proto-urban
conditions come bands of young and old drug addicts who have given up
their search for an ideal and live their sad nomadic lives from charity to theft.
This is the price, which, through an increasingly ungenerous public spirit,
we are refusing to pay.
The circle is nearing completion: our super-civilization is also and at the
same time a super-barbarism, the forms and protocols of which are slowly
becoming clear. The two compatible opposites, the increase in wild untamed
behaviour and the enlightened spirit that has accommodated this behaviour
and prepared its way, have combined to create a new hyper sensitivity
to fear. The enlightened are still unwilling to act against their vision of the
Utopian primitive and the wild love their wildness. As a result of this two
movements have arisen. On the one hand an increasingly paralysing technology of safety has been developed. On the other a boorish sect of quasi
military self-help cleansers has stood up to 'take matters in their own hand'.
This spiral of the enlightened wilderness has all the potential of blowing
up into a drama of ethnic recrimination requiring football-like loyalties and
aesthetically motivated violence. All this is set against an urban landscape
shaped according to the criteria of efficient logistics, scale enlargement and
economic volume. It has created a modern sublime of scale-less, coarsely
woven infrastructure with its plethora of dark, shadowy, empty places.
It is in this context that Shine 5.0 have done their subtle tweaking. They
do not wish to erase the product of modern economics, the landscape of
logistics, they wish to brace us against it, overcome it as it were. They wish
to tweak the city and tweak our minds, they wish to regain Aristotles' wonder-
ful definition of the city as a place where it is good to be, not by means of
an appeal to our conservatism and knee-jerkish attitude to everything that is
familiarly unfamiliar and by erasing the modern and supplanting it by the twee
and the comfortable, but by setting the comfortably familiar of the domestic
into the equally familiar setting of the alien. In this sense the wall-lessness of
their living room is interesting. *>> continues on page 148*

jaren zestig. Het antwoord op zich is dus misschien niet zo nieuw, maar wat wel nieuw is, is zowel de vraag als de context waarin het antwoord wordt gegeven. Daaraan ontleent het project zijn eigenheid. Deze open, muurloze stedelijke woonkamer is speciaal in het leven geroepen om iets te doen met ruimtes die door angst besmet zijn geraakt.

De wereld is veranderd. Niet alleen de voormalige koloniën, maar ook de Europese landen voelen nu steeds scherper wat het betekent als culturele interactie beperkt blijft tot etnische en religieuze tolerantie, en zich niet uit in positieve nieuwsgierigheid naar en plezier in elkaar. De problemen van een vaak unilateraal multiculturalisme, waarbij de ene persoon de andere tolereert en de ander die ene niet, spelen zich bij iedereen om de hoek af. Een van de vele producten van deze interne mislukking is de opkomst van de grootste mediaster aller tijden: de zelfvernietigende terrorist, die met zijn altruïstische daden van destructie en moord uit naam van zijn armzalige zaak de laatste restanten heeft gesloopt van dat maatschappelijke simulacrum, de morele grens. Tegelijkertijd worden we ruw geconfronteerd met de paradox van onze overgeciviliseerde samenleving. Vanuit ons verlichte sentiment, niet in de laatste plaats gevormd door de pedagogie van Jean-Jacques Rousseau, hebben we getracht van onze kinderen nobele wilden te maken: zelfstandige jongvolwassenen, zelfverzekerd, vol interesse en initiatief, mensen die op natuurlijke wijze kunnen leren van de natuur en het natuurlijke. We hebben daarbij behoorlijk wat succes gehad. Het resultaat wordt echter gecompliceerd door een groot aantal nobelen die eerder tot het wilde geneigd zijn, die vol verwarrende hormonen zitten en zich op het macho-model richten als ze zich een imago willen aanmeten en hun verlangen naar respect willen bevredigen. Uit dezelfde proto-urbane omstandigheden komen bendes jonge en oude drugsverslaafden, die het zoeken naar een ideaal hebben opgegeven en in hun trieste, nomadische bestaan leven van liefdadigheid en diefstal. Die prijs willen we in ons steeds hardvochtiger wordende publieke klimaat niet meer betalen.

De cirkel is bijna rond: onze superbeschaving is tegelijkertijd een superbarbarij waarvan de vormen en leefregels geleidelijk aan duidelijk worden. De twee verenigbare tegendelen – de toename van wild, ongetemd gedrag en de verlichte geest die dit gedrag heeft toegestaan en er de weg voor heeft geplaveid – hebben tezamen een nieuwe overgevoeligheid voor angst gecreëerd. De verlichten zijn nog altijd niet bereid hun visie op de utopische barbaar los te laten en de wilden genieten van hun wildheid. Hierdoor zijn twee bewegingen ontstaan. Enerzijds is er een steeds verlammender beveiligingstechnologie ontwikkeld, anderzijds is een boerse sekte van militaristische en eigenmachtige schoonmakers opgestaan die 'het recht in eigen hand neemt'. Er is alle kans dat deze spiraal van de verlichte wildernis uitgroeit tot een drama van etnisch ressentiment, inclusief de aan het milieu van de voetbalsupporters verwante loyaliteiten en esthetisch gemotiveerd geweld. Dit alles tegen de achtergrond van een stedelijk landschap gevormd door de criteria van efficiënte logistiek, schaalvergroting en economisch volume. Zo is een moderne grootsheid ontstaan van een schaalloze, grof geweven infrastructuur met een veelheid aan duistere, schimmige, lege plekken.

In deze context voert Shine 5.0 zijn subtiele fijnafstemmingen uit. De mensen van Shine 5.0 willen het product van de moderne economie, het landschap van de logistiek niet uitwissen, ze willen ons ertegen wapenen, het als het ware overstijgen. Ze willen de stad fijn afstemmen en onze geest ook, ze willen terug naar Aristoteles' prachtige definitie van de stad als 'een plaats waar het prettig toeven is', niet via een beroep op ons conservatisme en onze geconditioneerde houding tegenover alles wat ons als onbekend vertrouwd is, en ook niet door het moderne uit te wissen en te vervangen door het sentiment en het vertrouwde, maar door het prettig vertrouwde van het huiselijke onder te brengen in de al even vertrouwde omgeving van het onbekende. In dat opzicht is de muurloosheid van hun woonkamer interessant. *>> vervolg op pagina 149*

Fear as Imagination

I move through the city in a particular manner because I have constructed a particular psycological reality made of distinctions, or powerful fantasies of distinctions, based on, not rational intentions or cognitions, but rationalizations of unconscious desires.[1]

Contrary to popular belief, agoraphobia is not a fear of open space antonymous with claustrophobia, but a fear of both urban space and its events. The original German term *Platzangst* comes closer to defining the phobia – fear of (public) space. Agoraphobia developed as a symptom of urban society: without the city it cannot exist.
Tweak is a study of fear in the non-private space – in movement between places of relative safety – in the west of Rotterdam. At the heart of this project is the premise that 'fear is imagination. Fear is not created by space, rather the space is redefined by mental sets associated with a stereotype. The dark empty street, the unpopulated canalside path, the fluorescent railway underpass: these are the blank canvasses where the mind can animate imaginary negative events.'[2] Tweak searches for the stimuli that motivate the projection of negative images on the space, aiming to balance fear with reality instead of imaginary projections. Armed with a new kind of understanding of the relationship between the place and its inhabitants, the conditions that stimulate the projection of the imaginary fear landscape may be manipulated – *tweaked*. (1-3)

1 Agoraphobia: de verlaten straat
1 Agoraphobia: the deserted street

2 Sociale bezigheid: drie meisjes
praten buiten voor hun huis
2 Social occupation: three girls
chatting outside their home

1 Dora Epstein, 'Abject Terror: A Story of Fear, Sex, and Architecture',
in: Nan Ellin (ed.), *Architecture of Fear*, New York 1997, p. 133.
2 Excerpt from the initial statement of Shine 5.0 in their application
for Groepsportretten 2004.

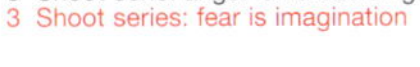
3 Shoot serie: angst is verbeelding
3 Shoot series: fear is imagination

De verbeelding van de angst

*Ik beweeg me op een bepaalde manier
door de stad omdat ik een bepaalde
psychische werkelijkheid heb gecreëerd
van distincties, of levendige fantasieën
van distincties, die niet gebaseerd
zijn op rationele intenties of cognities,
maar op rationalisaties van onbewuste
wensen.*[1]

In tegenstelling tot wat de meeste
mensen denken is agorafobie, de angst
voor open ruimte, niet het tegenovergestelde van claustrofobie, het is de
angst voor zowel de stedelijke ruimte
als de gebeurtenissen die er plaats-
vinden. De oorspronkelijke Duitse term
Platzangst definieert deze fobie beter:
angst voor (publieke) ruimte. Agorafobie
is ontstaan als symptoom van de
stedelijke samenleving: zonder de
stad kan ze niet bestaan.
Tweak is een onderzoek naar angst in
de niet-private ruimte, in de beweging
van de ene relatief veilige plaats naar
de andere. Kern van het onderzoek
vormt de premisse 'angst isverbeelding.
Angst wordt niet veroorzaakt door ruimte, het is de ruimte die wordt geherinterpreteerd aan de hand van mentale
beelden die vergezeld gaan van een
stereotype. De donkere, verlaten straat,
het eenzame weggetje langs het water,
de helverlichte tunnel onder het spoor:
stuk voor stuk lege schermen waarop
de menselijke geest denkbeeldige
negatieve gebeurtenissen kan projecteren'.[2] *Tweak* gaat op zoek naar wat
mensen prikkelt om negatieve beelden
op het scherm te projecteren, met als
doel angst in evenwicht te brengen
met de werkelijkheid, in plaats van met
geprojecteerde negatieve verbeelding.
Met gebruikmaking van een nieuw
inzicht in de relatie tussen de plek
en de mensen die er wonen, zijn de
omstandigheden die prikkelen tot
de projectie van een imaginair angst-
landschap, te manipuleren, in het
Engels gezegd: te *tweaken*. (1-3)

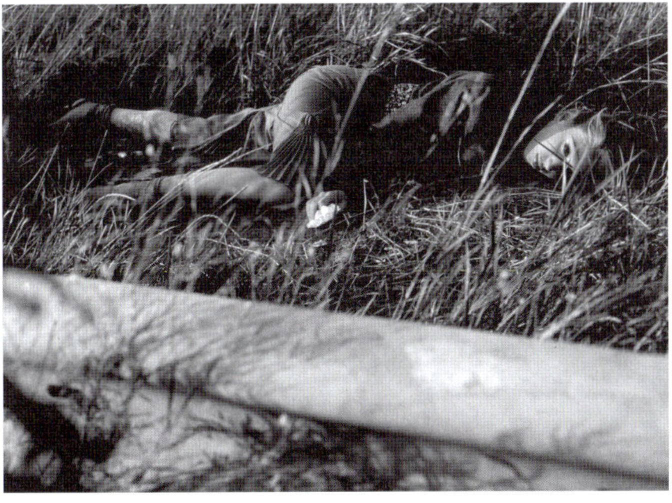

1 Dora Epstein, 'Abject Terror. A Story of Fear, Sex and Architecture',
  in: Nan Ellin (red.), *Architecture of Fear*, New York 1997, p. 134.
2 Passage uit de eerste uiteenzetting van Shine 5.0
  in hun aanmelding voor *Groepsportretten* 2004.

Enviromotional Geography: Mapping and Results

We map the cityscapes into placesI will go and places I will not, [narrating] our cartographies of avoidance, our fearing.[3]

Marconiplein is a transport hub (and little more) in the west of Rotterdam, and as such a concentration of urban movement. It also offers a diverse range of communities, in the five surrounding suburbs, Spangen and Bospolder (both classified as 'problem areas' areas by the Municipality of Rotterdam); Tussendijken (a 'threatened area'); and the Witte Dorp and Nieuwe Mathenesse ('attention areas', the latter including Rotterdam's 'Toleration Zone' for prostitution and drugs). Interestingly almost all of the areas have improved since 2003, moving up one category in the 'safety index'.[4] However, although fear increases steadily as actual safety decreases, the relationship becomes disproportional as the area begins to recover. The image of fear has already become part of the area's identity. (4)

Signals of fear, both human and environmental, are recorded, visualized and overlapped. The maps produced are a sensual, subjective abstraction combining visual, graphic and sensory elements.

The text 'Fear of Crime',[5] highlights *incivilities* as the indicators of an increase in crime in the environment and therefore possible stimuli for fear. The environmental mapping shows, for example, that graffiti proliferates in Spangen, Tussendijken and Bospolder, but neither in Marconiplein itself nor in the Witte Dorp; there are concentrations of rubbish in Tussendijken and Spangen, particularly on Mathenesserdijk; there are two epicentres for broken and boarded-up windows, one between Mathensserdijk and Mathenesserweg in Spangen, and the other on the east of the site in Bospolder. (5)

4 Observatie van stedelijke activiteit: Marconiplein
4 Urban movement observation: Marconiplein

3 Dora Epstein, 'Abject Terror: A Story of Fear, Sex, and Architecture', in: Nan Ellin (ed.), Architecture of Fear, New York 1997, p. 133.
4 Safety Index 2004, Municipality of Rotterdam, May 2004, www.rotterdam.nl/veilig.
5 Fear of Crime, John Howard Society of Alberta, 1999, see: www.johnhoward.ab.ca/PUB/C49.htm.

4 Het beeld van de angst als onderdeel
  van de identiteit van het gebied
4 The image of fear as part of the
  area's identity

Omgeving en gevoelens in kaart: onderzoek en resultaten

We classificeren het landschap van de stad naar plaatsen waar je wel komt en plaatsen waar je niet komt en maken zo een kaart van de dingen die we vermijden, van onze angsten.[3]

Het Marconiplein is niet veel meer dan een verkeersknooppunt in Rotterdam-West, met veel stedelijke activiteit. Er wonen uiteenlopende gemeenschappen in de vijf omringende wijken Spangen, Bospolder (door de gemeente Rotterdam geclassificeerd als 'probleemgebieden'), Tussendijken ('bedreigd gebied') en het Witte Dorp en Nieuw Mathenesse ('aandachtsgebieden', met in laatstgenoemde wijk Rotterdams 'gedoogzone' voor prostitutie en drugs). Het interessante is dat sinds 2003 in bijna al deze gebieden verbetering is opgetreden; ze scoren allemaal één categorie hoger op de 'veiligheidsindex'.[4] Maar hoewel angst geleidelijk toeneemt naarmate de feitelijke veiligheid afneemt, raakt dit verband zoek naarmate het gebied zich herstelt. Het beeld van de angst is al onderdeel geworden van de identiteit van het gebied. (4)

Angstsignalen, van anderen en uit de omgeving, worden geregistreerd, gevisualiseerd en gecombineerd. De plattegronden die hieruit ontstaan, zijn zintuiglijke, subjectieve abstracties met een mengeling van visuele, grafische en sensorische elementen.

In de tekst 'Fear of Crime'[5] wordt nadruk gelegd op *overlast* en *vervuiling* als indicatoren van een toename van criminaliteit in de omgeving en dus als mogelijke prikkels van angst. Onderzoek naar omgevingsfactoren wijst bijvoorbeeld uit dat er in Spangen, Tussendijken en Bospolder veel graffiti is, maar niet op het Marconiplein zelf, noch in het Witte Dorp. Op bepaalde plekken in Tussendijken en Spangen ligt veel rotzooi op straat, vooral op de Mathenesserdijk. Er zijn twee epicentra

3 Dora Epstein, 'Abject Terror. A Story of Fear, Sex and Architecture',
  in: Nan Ellin (red.), Architecture of Fear, New York 1997, p. 134.
4 Veiligheidsindex 2004, Gemeente Rotterdam, mei 2004,
  www.rotterdam.nl/veilig
5 Fear of Crime, John Howard Society of Alberta, 1999,
  www.johnhoward.ab.ca/PUB/C49.htm

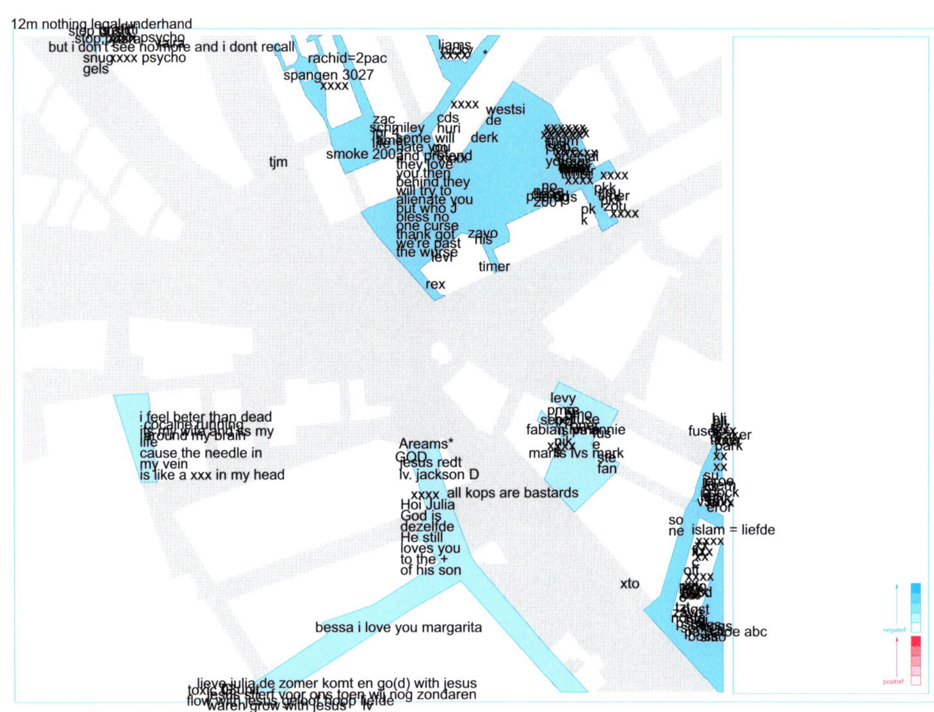

5 Environmental mapping:
   incivilities: graffiti

5 Environmental mapping:
   incivilities: Graffiti

6 Emotional mapping: interview 26:
   moeder haalt haar kinderen op van
   school

6 Emotional mapping: interview 26:
   mother picking up her children from
   school

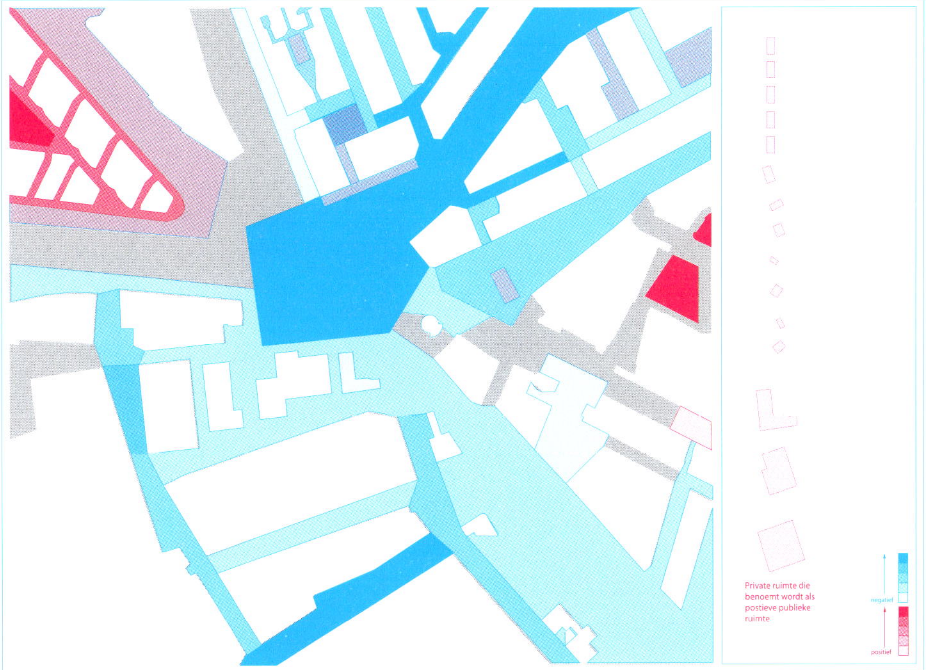

7 Emotional mapping:
   interview 6a: buren
7 Emotional mapping:
   interview 6a: neighbours

8 Enviromotional geography 1:
   overlappen van onveilige zones
   en beweging volgens individuele
   beleving
8 Enviromotional geography 1:
   overlap of individual spatial fear
   zones and movements

The emotional mapping, generated through interviews and observation, attempts to understand the conscious and subconscious cartographies of fear in the local community. (6-7) Clearly the zones that denote a positive image related to pleasure (shown in percentages of magenta) – such as the Witte Dorp or the many playgrounds – rarely coincide with the incivility concentrations. Residents of the Witte Dorp repeatedly cite their own neighbourhood as a safe and pleasurable space, because of family and neighbourly familiarity,[6] and because of the availability of collective and private space (pedestrian streets, gardens, a central square and community centre).[7] Playgrounds receive similar good reports,[8] presumably due to their liveliness and collectiveness. (8)

In contrast, zones that denote negative emotional space (shown in percentages of cyan) correlate closely with concentrations of incivilities. For example, Mathenesserdijk emerges as a high concentration zone for litter, graffiti and boarded-up windows. This street becomes deeper cyan due to its repeated citation as a 'place I would not walk due to fear'. In the morning, 'when the junks are coming back from Marconiplein',[9] Mrs. 6a's movements are limited to such an extent that she will not walk the 50 m to Marconiplein to use public transport and consequently limits her normal daily footprint to the area between her front door and her car. In the evening, dealers inhabit Mathenesserweg in avoidance of the surveillance cameras in Marconiplein.[10] Keileweg, the infamous tolerance zone, also appears in cyan, although it is not cited often because it is not in the normal daily space of the residents of the test-area. Once, Keileweg and the surroundings are cited as a positive space, in this case by someone who works there and considers it familiar territory.[11] (9-10)

9 Incivility: graffiti, beplate vensters
   en afbraakpanden
9 Incivility: graffiti, boarded-up
   windows and derelict buildings

6 Interview a12.
7 Interview a12.
8 Interview a07, a30, c20.
9 Interview c6a.
10 Interview c05.
11 Interview a04.

met veel kapotte en dichtgespijkerde ramen, één tussen de Mathenesserdijk en de Mathenesserweg in Spangen en het andere ten oosten van het Marconiplein in Bospolder. (5)

Aan de hand van interviews en observaties is een gevoelsplattegrond samengesteld, een poging de bewuste en onderbewuste angsten van mensen in de plaatselijke gemeenschap in kaart te brengen. (6-7) De zones die een positief beeld oproepen in termen van woongenot (aangegeven in percentages in hardroze), zoals het Witte Dorp of de vele speelplaatsen, vallen zelden samen met concentraties van overlast en vervuiling. Inwoners van het Witte Dorp noemen herhaaldelijk hun eigen buurt als een veilige en prettige omgeving, omdat ze familie en kennissen in de buurt hebben,[6] en omdat er voldoende gemeenschappelijke en private ruimte beschikbaar is (voetgangersgebieden, tuinen, een centraal plein en een buurthuis).[7] Ook speelplaatsen worden vaak positief beoordeeld,[8] waarschijnlijk dankzij aspecten als levendigheid en gemeenschappelijkheid. (8)

De zones die geïndiceerd zijn als negatieve emotionele ruimte (aangegeven in percentages in cyaanblauw) overlappen daarentegen wel met de concentraties overlast en vervuiling. Op de Mathenesserdijk zie je bijvoorbeeld veel afval, graffiti en dichtgespijkerde ramen. Omdat deze straat herhaaldelijk genoemd wordt als 'plaats waar ik uit angst niet zou komen', is hij aangegeven in donker cyaanblauw. 's Morgens, 'als de junkies terugkeren van het Marconiplein'[9], weigert de bewoonster van 6a zelfs de vijftig meter tot het Marconiplein te overbruggen om de bus of de tram te pakken en beperkt ze haar gang tot het gedeelte tussen de voordeur en haar auto. 's Avonds wordt de Mathenesserweg bevolkt door dealers, die uit het zicht proberen te blijven van de surveillancecamera's op het Marconiplein.[10] De Keileweg, de beruchte gedoogzone, is ook in

6 Interview a12.
7 Interview a12.
8 Interview a07, a30, c20.
9 Interview c6a.
10 Interview c05.

10 Angst: Beweging, collectieve
    zones, tijd en activiteit
10 Fear: Mobility, collective zones,
    time and activity

1000 - 1400
0600 - 1000
2200 - 0600
2200
2000 2100 2200 2300 2400 0100 0200 0300 0400 0500 0600 0700 0800 0900 1000

An additional layer of information is introduced in the *Repeated Words* from 50 interviews. The most common repetitions concern stereotypes (junks), positive and negative locations (playgrounds, Mathenesserweg) racial and social prejudices (not safe for whites), the unsolicited citing of fearful times (Mathenesserweg in the morning when the Tolerance Zone closes), and serial misinterpretation of boundaries (citing 'my house' or 'my garden' as an example of positive public space). (11-12)

Although openly unscientific, stories and observations revealed strong and often unexpected overlaps, patterns and concentrations. These results were used to formulate four loose but provocative statements about the phenomenon of fear and its roots in the Marconiplein Area.

11 Interviews: op waarneming
gebaseerde informatie van
bewoners
11 Interviews: inhabitant's
perceptual information

11 Interview: op waarneming
gebaseerde informatie van
bewoners
11 Interview: inhabitant's
perceptual information

12 Gevonden voorwerpen
12 Found objects

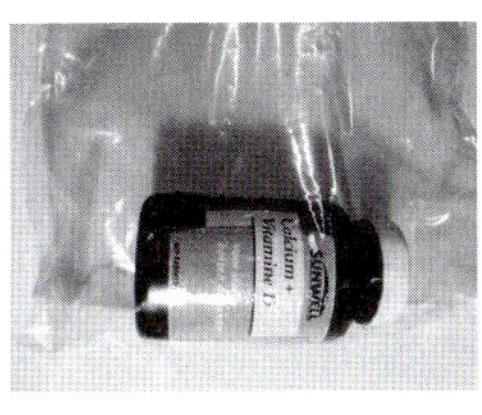

blauw aangegeven hoewel het gebied niet vaak wordt genoemd, omdat het geen deel uitmaakt van de normale leefomgeving van de bewoners van het onderzoeksgebied. Eén keer is de Keileweg en omgeving aangemerkt als positieve ruimte, in dit geval door iemand die er werkt en het beschouwt als vertrouwd gebied.[11] (9-10)

'Herhaalde Woorden' uit vijftig interviews biedt een extra laag informatie. De meest voorkomende herhalingen betreffen stereotypen (junks), positieve en negatieve locaties (speelplaatsen, Mathenesserweg), raciale en sociale vooroordelen (niet veilig voor blanken), spontaan vertelde voorbeelden van angstige momenten (de Mathenesserweg in de ochtend als de gedoogzone dichtgaat). Ook blijken de grenzen herhaaldelijk verkeerd te worden geïnterpreteerd ('mijn huis' of 'mijn tuin' worden aangehaald als voorbeelden van positieve openbare ruimte). (11-12)

Hoewel het onderzoek bewust niet wetenschappelijk is, vertoonden de verhalen en observaties sterke en vaak onverwachte overeenkomsten, patronen en concentraties. Op basis van de resultaten zijn vier losse doch provocerende stellingen geformuleerd over het verschijnsel angst en de oorzaken ervan in het gebied rondom het Marconiplein.

11 Interview a04.

Loosely Provocative: Conclusions

Fear is mobile, moving through the city with movements of particular people, lights, cameras...
Fear is the unfamiliar: safety and comfort is in the private domain.
Fear is event: it is not directly related to space, rather space is a consequence of the event.
Fear is social disconnection, the consequence of overlapping but disintegrated social networks.

Tweaking: Interventions

What is the modern house, if not a cherished retreat from agoraphobia.[12]

On Matthenesserdijk, in Spangen, what seems to the outsider to be a vibrant and diverse street culture is perceived by the locals as a culture of spatial poverty and fear: children cannot play computer games all day, they have to play outside, but since the residents have no private outdoor space, they have to play on the street or in the playground. These public spaces are often too dangerous for the children alone, so the parents or other family members must accompany them. What results is an overlapping but disconnected and suspicious society. In *Tweak*, the information gathered in the research phase is used to make a number of related interventions that *tweak* – that is change the context in a small but important way – in order to reprogram the stimuli for fear. In particular, the issues of mobility, familiarity, event and social disconnection are confronted as *tweak*able phenomena. (13)

Tweak consists of a number of 2d interventions – a typical and personalized local floor plan (taken from Brinkman's Van Lennepblok in Spangen) at scale 1:1 painted in the public space with reservations for major components such as the sofa, the television, the

13 Huiskamer op straat
13 Public living room series

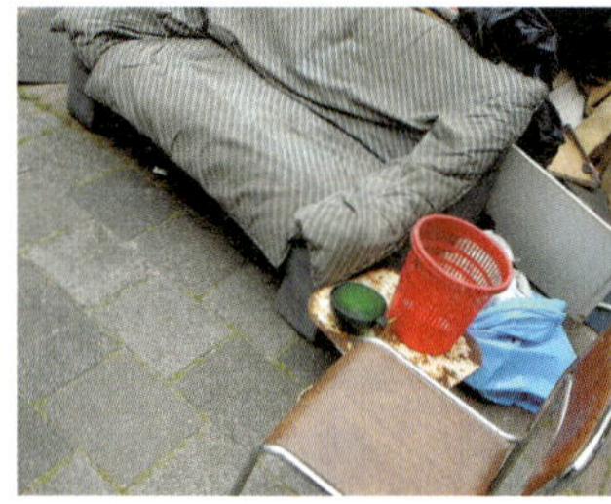

12 Anthony Vidler, *Warped Space, Art, Architecture, and Anxiety in Modern Culture*, Cambridge, Mass. 2000, p. 148.

Losjes provocerend: conclusies

Angst is mobiel, hij verplaatst zich door de stad in het voetspoor van bepaalde mensen, lichten, camera's…
Angst is het onbekende: veiligheid en comfort vind je in de private ruimte.
Angst is gebeurtenis: angst is niet direct gerelateerd aan ruimte, de ruimte is eerder het gevolg van de gebeurtenis.
Angst is gebrek aan sociale samenhang, het gevolg van elkaar overlappende maar uiteenvallende sociale netwerken.

Tweaking: interventies

Wat heb je aan een modern huis, als je er niet heerlijk in kunt wegvluchten.[12]

De voor buitenstaanders bruisende en afwisselende straatcultuur van de Mathenesserdijk in Spangen wordt door de bewoners beleefd als een cultuur van ruimtegebrek en angst. De kinderen kunnen niet de hele dag computerspelletjes doen, ze moeten buitenspelen, maar aangezien de bewoners geen private buitenruimte hebben, moeten de kinderen op straat spelen of op de speelplaats. Omdat deze publieke ruimtes vaak te gevaarlijk zijn voor kinderen alleen, moeten hun ouders of andere familieleden mee. Het resultaat is een overlappende, maar onsamenhangende en achterdochtige samenleving.
De informatie uit de onderzoeksfase van *Tweak* is gebruikt om een aantal samenhangende ingrepen te doen die kleine maar belangrijke wijzigingen (*tweaks*) aanbrengen in de context, om zo de prikkels van de angst te herprogrammeren. Vooral zaken als mobiliteit, vertrouwdheid, gebeurtenissen en gebrek aan sociale samenhang worden gezien als *tweakable*, oftewel manipuleerbare verschijnselen. (13)

Tweak bestaat uit een aantal tweedimensionale ingrepen: op verschillende plaatsen in de openbare ruimte is op schaal 1:1 de plattegrond afgetekend van een typische woonkamer in de

12 Anthony Vidler, *Warped Space, Art, Architecture,
and Anxiety in Modern Culture*, Massachusetts Institute
of Technology, Cambridge 2000, p. 148.

photograph frames – and one mobile
3d intervention (the components themselves, distorted in scale according to
their role) which moves between the
sites. The sites and the cycle of these
'LivingRooms' are derived from the
concentration zones highlighted in the
mapping phase. (14)

When a *Living*Room is in place,
local and personal stories can be
recorded on the typewriter. Pictures,
ornaments, and even additional furni-
ture can be brought and arranged in
the space. A film projected onto the
oversized television shows the *Living*-
Room in its previous location and the
interaction and stories of the near
neighbours. When the *Living*Room is
gone, the floor markings remain. As
the *Living*Room travels, it undergoes
a metamorphosis; although returning
to the same footprint each time, the
collected identities and stories from
the community gradually transform the
*Living*Room into a collective private
space. Negative local objects are
transformed to create a mobile space
where the local identities can be shared
and stereotypes can be gently reprogrammed. (15-16)

A place is created, an imaginary room,
with associations of safety and control.
Here, an indirect communication is
enabled between anonymous near-
neighbours who fear each other through
imagination and ignorance. The time-
delayed presentation of the *Living*Room
in its previous location is essential in
connecting with other sites, physically
close, but mentally distant. As time
goes by, the *Living*Room transforms
from a representation of *someone's*
living room into a community living
room where identities are contributed
by the collective and stereotypes can
be reprogrammed.

As the twenty-first century inhabitants of the city retreat into the private
domain, the leftover public realm is
becoming increasingly anonymous

14 Woonkamer serie
14 Livingroom sequence

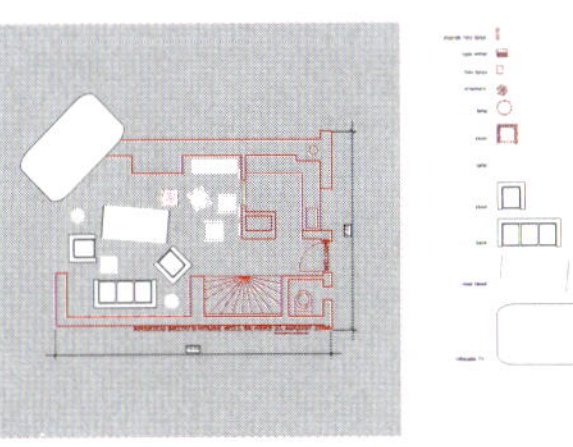
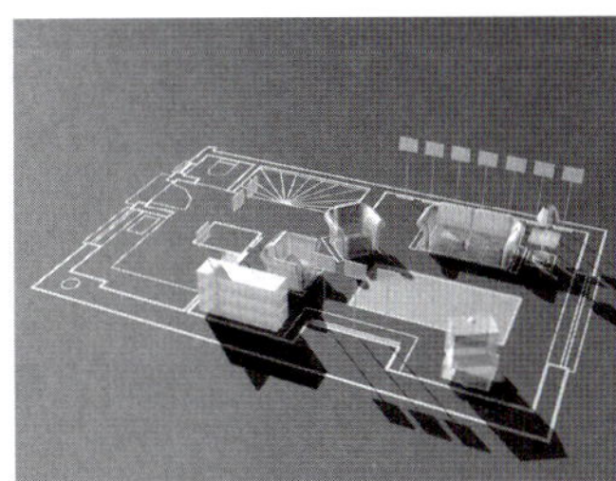

14 Woonkamer onderdelen
14 Livingroom components

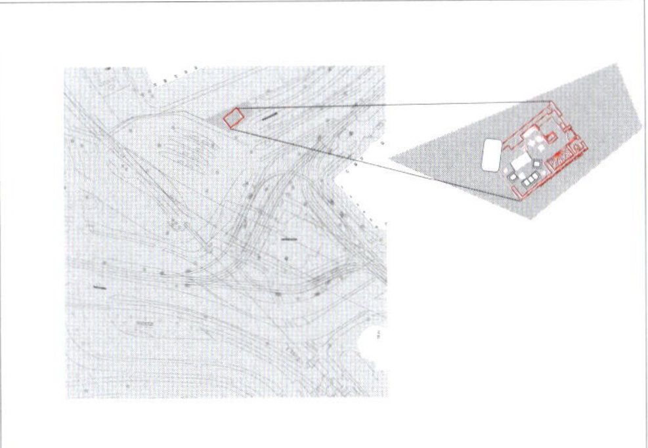

14 Woonkamer locatie: Marconiplein /
    Mathenesserdijk
14 Livingroom location: Marconiplein /
    Mathenesserdijk

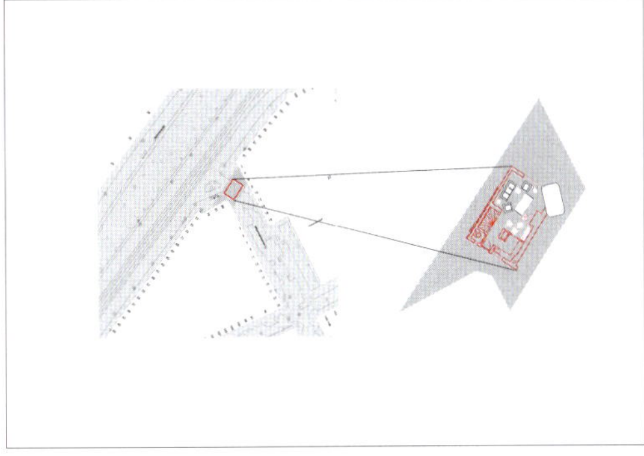

14 Woonkamer locatie:
    Willem Beukelszstraat
14 Livingroom locations:
    Willem Beukelszstraat

buurt (een werkelijk bestaande woonkamer uit het woongebouw van Brinkman aan de Van Lennepstraat in Spangen), met open plekken voor de belangrijkste elementen zoals de bank, de televisie, fotolijstjes; daarnaast is er een mobiele driedimensionale ingreep (het 'meubilair' zelf, waarvan elk element is uitgevoerd op een schaal die beantwoordt aan zijn rol) die rondreist tussen de verschillende open plekken. De plekken en de rondgang van deze 'Woonkamers' zijn ontleend aan de concentratiezones uit de 'karteringsfase'. (14)

Zodra een *Woon*kamer is ingericht, kan men er lokale en persoonlijke verhalen optekenen aan de typemachine. Men kan foto's, decoraties en zelfs extra meubilair meenemen en in de ruimte plaatsen. Op de buitenproportioneel grote televisie wordt een film vertoond die de *Woon*kamer laat zien op de vorige plek waar ze stond en de interactie en verhalen van de bijna-buren. Als de *Woon*kamer weg is, blijven de markeringen op de grond zichtbaar. De *Woon*kamer verandert terwijl ze rondreist; hoewel ze telkens terugkeert naar de oorspronkelijke plattegrond, verzamelt de *Woon*kamer identiteiten en verhalen uit de gemeenschap en transformeert zo geleidelijk tot een gemeenschappelijke private ruimte. Negatieve lokale objecten worden getransformeerd tot een mobiele ruimte waar lokale identiteiten gedeeld, en stereotypen voorzichtig opnieuw geprogrammeerd kunnen worden. (15-16)

Er is een plek ontstaan, een denkbeeldige kamer, waar men zich veilig en gecontroleerd voelt. Indirecte communicatie is hier mogelijk tussen anonieme bijna-buren die elkaar vrezen door inbeelding en onwetendheid.
Het achteraf vertonen van beelden van de *Woon*kamer in haar vorige positie is essentieel om verbanden te leggen met andere plaatsen die fysiek nabij zijn maar mentaal veraf. Met het verstrijken van de tijd verandert de *Woon*kamer

Tweak   Shine 5.0

and vacant. A vicious circle is created, where the private space grows internally or virtually while the shrinking public space is colonized by a sub-culture, whole neighbourhoods decline, and municipalities seek remedy through demolition and new-build.

The *Living*Room questions the role of public space in integrating communities as bounded space takes over and produces fear in and of the un-private space. It is a comment on the lack of private outdoor space and the associated fear of public space. It reacts to the event-like and mobile nature of fear in the public domain and offers the community the chance to begin to interact with each other in a pseudo-safe environment so that their fears are brought into line with reality. Finally, imaginations can project positive images back onto the environment that stimulates them. (17-18)

15 Op locatie: Willem Beukelszstraat:
YOUR LIVING ROOM WILL
BE HERE
15 On location: Willem Beukelszstraat:
YOUR LIVING ROOM WILL
BE HERE

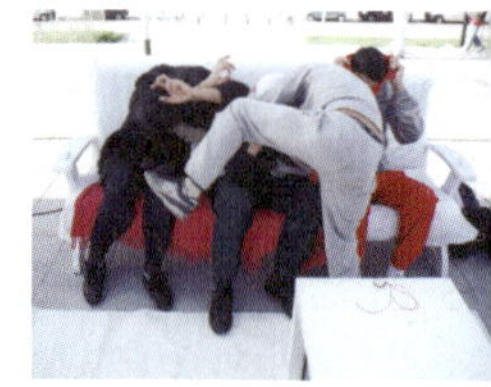

16 Woonkamer op locatie:
   Marconiplein / Mathenesserdijk
16 Livingroom on location:
   Marconiplein / Mathenesserdijk

van een representatie van *iemands* woonkamer in een gemeenschappelijke woonkamer waar het collectief identiteiten bijdraagt en waar stereotypen geherprogrammeerd kunnen worden.

De bewoners van de eenentwintigste eeuw trekken zich terug in hun private domein, waardoor het publieke domein dat achterblijft alsmaar anoniemer en leger wordt. Er is een vicieuze cirkel ontstaan: de private ruimte neemt toe, intern of virtueel, terwijl de inkrimpende openbare ruimte wordt gekoloniseerd door een subcultuur, hele wijken in verval raken en gemeentebesturen hun heil zoeken in kaalslag en nieuwbouw.

De *Woon*Kamer stelt vragen bij de rol van de openbare ruimte in de integratie van gemeenschappen, waarbij de omsloten ruimte de overhand krijgt en angst in en voor de niet-private ruimte in de hand werkt. Het project is een commentaar op het gebrek aan private buitenruimte en de daarmee verbonden angst voor de publieke ruimte. Het is een reactie op de angst in het publieke domein, die naar zijn aard gebonden is aan gebeurtenissen en mobiliteit, en het biedt de buurtbewoners een mogelijkheid om met elkaar in contact te komen in een pseudo-veilige omgeving, die hun angsten in evenwicht brengt met de werkelijkheid. Tot slot kan de verbeelding ook positieve beelden terugprojecteren op de omgeving waardoor ze is geprikkeld. (17-18)

Loosely provocative:
reaction on conclusions

Event displaces the perceived fear
helping to reclaim the territory for
the neighbourhood.

Keep it simple, listen, use the 'street
language' and customs of the area
and feedback into the community.

Create social connection through
event to make the street a home of
the community.

Make non buildings and areas into
places of reflective focus by event
or tweak.

17 Livingroom op locatie:
    Marconiplein / Mathenesserdijk
17 Livingroom on location:
    Marconiplein / Mathenesserdijk

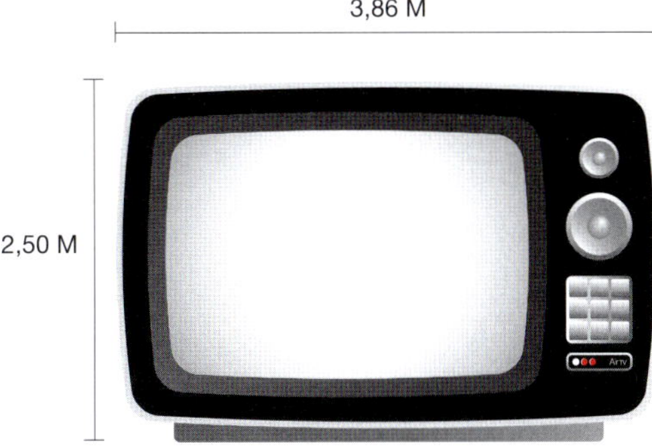

Losjes provocerend:
reacties op eerdere conclusies

De gebeurtenis verdringt het gevoel van angst en geeft het territorium terug aan de buurt.

Hou het eenvoudig, luister, spreek de 'taal van de straat' en hou je aan de manier waarop mensen hier met elkaar omgaan, en zorg dat je terugkoppelt naar de gemeenschap.

Zorg voor sociale samenhang door middel van de gebeurtenis; maak zo van de straat een thuis voor de gemeenschap.

Maak van non-gebouwen en -plaatsen plekken van reflectie door gebeurtenissen of tweaks.

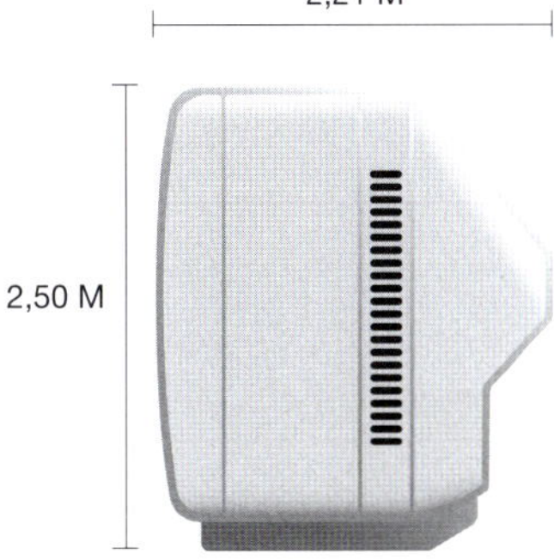

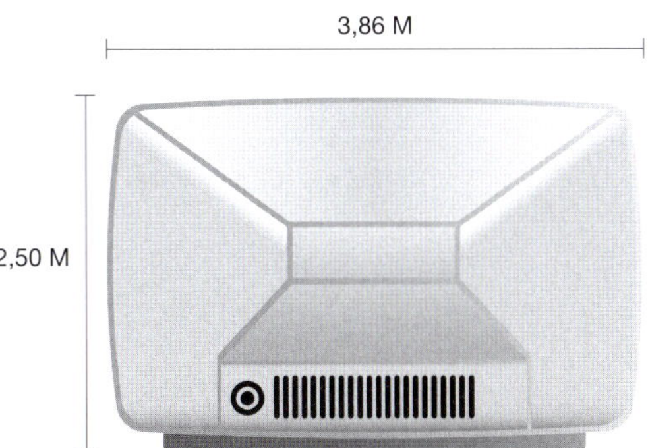

18 Ontwerp schets voor
    opblaasbare TV
18 Prototype design for
    inflatable TV

<< continued from page 124 / Jacob Voorthuis, The Enlightened Wilderness

Fear is a curious concept in that, more than any other word, it is defined within the blind parameters of cultural and perhaps biological prejudice. For the purposes of this essay we need not dig too deep, but it would be useful to define fear not in terms of the catalogue of different possible fears and phobias such as agora-, claustro- and assymmetrophobia. Nor do we need to look at the theatre of repression and the metaphysical psychology of Freud. What we need to do is to look at fear as a praxis, as an activity rather than a feeling. To define fear pragmatically we have to have an idea of what people with fear do. Fear as an activity is similar to the Aristotelian theoria, a form of exercise whereby one's image and knowledge of the world is consciously brought forth, contemplated, rehearsed and its internal structure made subject to a rigorous critique. In this sense fear, as a dialectical activity upon a flood of images and affects in the mind is an act of continuous portraiture. The connection between fear and mobility that the Shine group have seized upon is cogent. Movement and fear are complementary, even if that move-ment is realized only as a wish. Fear is an act of portraiture, in which modalities are given concrete form and rehearsed in provisionally prepared responses as the familiar landscape of fear unfolds itself to us as we walk. The repainting of the object of fear is continuous; it is like the Mona Lisa, continually reworked and continually reinterpreted: now the portrait of a middleclass housewife, then the possible portrait of Leonardo himself. . . Fear is a portrait which is continually reworked up until the moment of the artist's death. It forms part of his legacy; in the lines of his face he leaves behind him the most complete portrait of his fears. This portrait, apart from being an instrument of self-revelation, is also magical in Collingwood's classic sense. A person's portrait of fear is not a disinterested product; it is primarily a useful thing, or at least a thing that can be acted upon. It creates possible situations in the mind and prepares appropriate gestures and actions. The person walking through the city moves through an urban gallery containing the portraits of his mind. This portrait could be mere myth, it could represent reality, but it is always virtual in that it is always an image of a structured possibility: an imaginary monster related to the real and to the fantastic through the infinite world of possibility.

Fear, like the personal law-enforcement that is violence, is an extraordinary powerful generator of urban usage and form precisely through this continuous act of life-like and, at the same time fantastic portraiture of modalities. The possible does not like to discriminate too rigidly between fantasy and the actual. Although such an act of portraiture is highly indi-vidual, it nevertheless has the power to shape the city. The city is a product of individual movements multiplied by the numbers and desires of its population. This, within the parameters of daily life, creates patterns of usage and avoidance which in turn influence the art of political prioritization and thus designed urban form.

Imagine now the grafting of the comfortably familiar, such as a simulated living room, on to the territory of the alien. The walls of this incongruous simulacrum are merely defined by the inscriptions on the ground. As such the non-existent walls become overwhelming views of the dull-grey back-ground of the modern sublime, much like the walls of the Sala dei giganti by Giambattista Bologna in the Palazzo del Te in Mantua: space disfigured by the overwhelming omnirama of destruction. The non-walls of the simulated living room are full cinematographic frescoes of a dreary background upon which visitors can superimpose their own personal portraits of potential at leisure. The walls are virtual divisions, cinematographic and not physical. They define a model space, a space full of uncontrollable possibilities. The living room will simulate the haven, will contain the instruments of a quiet

<< vervolg van pagina 125 / Jacob Voorthuis, De verlichte wildernis

Angst is in zoverre een merkwaardig begrip dat het meer dan enig ander
woord wordt bepaald door de blinde parameters van cultureel en wellicht
biologisch vooroordeel. Voor dit artikel hoeven we niet al te diep te spitten,
maar het zou nuttig te zijn om angst eens niet te definiëren in termen van
het scala aan mogelijke angsten en fobieën, zoals agora-, claustro- en
asymmetrofobie. Evenmin hoeven we het theater van de verdringing en de
metafysische psychologie van Freud erbij te halen. Wat we wel moeten doen
is kijken naar angst als praxis, niet als gevoel maar als activiteit. Om angst in
pragmatische zin te kunnen definiëren, moeten we een idee hebben van wat
bange mensen doen. Angst als activiteit is te vergelijken met de aristotelische
theoria, een soort oefening waarbij iemands beeld en kennis van de wereld
bewust naar buiten worden gebracht, overdacht en gerepeteerd en in hun
interne structuur onderworpen aan rigoureuze kritiek. Zo bezien is angst,
opgevat als een dialectische activiteit in reactie op een vloed aan beelden
en affecten in de geest, een daad van voortdurende portrettering.
Het verband tussen angst en mobiliteit waar de Shine-groep zich op con-
centreert, is overtuigend. Beweging en angst zijn complementair, zelfs als
die beweging alleen als wens bestaat. Angst is een daad van portrettering:
modaliteiten krijgen concreet vorm in provisorisch voorbereide reacties op
het vertrouwde landschap van de angst dat zich onderweg aan ons open-
baart. Het object van de angst wordt steeds opnieuw geschilderd; het is als
de *Mona Lisa* die voortdurend wordt bijgewerkt en geherinterpreteerd: nu
eens als het portret van een doorsneehuisvrouw, dan weer als een mogelijk
portret van Leonardo zelf… Angst is een portret dat voortdurend wordt
bijgewerkt tot het moment dat de kunstenaar sterft. Het is onderdeel van
zijn erfenis; in de lijnen van zijn gezicht laat hij het meest complete portret
van zijn angsten na. Dit portret is niet alleen een instrument van zelfopen-
baring, het is ook magisch in de klassieke zin volgens Collingwood. Iemands
angstportret is geen neutraal product; het is primair iets wat handig is in het
gebruik, althans het gedrag. Het creëert mogelijke situaties in de geest en
bereidt passende gebaren en handelingen voor. De persoon die door de stad
loopt, beweegt zich door een stadsgalerie waar zijn geestelijke portretten
te zien zijn. Zo'n portret kan louter mythe zijn of het zou de realiteit kunnen
verbeelden, maar het is altijd virtueel in de zin dat het altíjd een beeld is
van een geconstrueerde mogelijkheid; een denkbeeldig monster dat via de
onbegrensde wereld van het mogelijke zowel verbonden is met de werkelijkheid als met de fantasie.
Net als de persoonlijke wetshandhaving is angst een uitzonderlijk krachtige generator van het gebruik en de vorm van de stad, juist vanwege dit
voortdurend levensecht of fantastisch portretteren van mogelijkheden.
Het mogelijke maakt liever geen al te rigide onderscheid tussen fantasie
en werkelijkheid. Dit portretteren mag dan een hoogst individuele daad zijn,
het heeft wel het vermogen de stad vorm te geven. De stad is een product
van individuele bewegingen, vermenigvuldigd met de aantallen en de
verlangens van zijn inwoners. Dit creëert binnen de begrenzingen van het
dagelijks leven patronen van gebruik en ontwijking, die op hun beurt de
kunst van het stellen van politieke prioriteiten bepalen en zodoende de
ontworpen stedelijke vorm.
Stel u nu voor dat het prettig vertrouwde, bijvoorbeeld een nagebouwde woonkamer, wordt overgeplant naar het domein van het onbekende. De muren
van dit ongerijmde schijnbeeld worden slechts aangegeven door lijnen op
de grond. Zo worden de niet-bestaande muren overweldigende uitzichten op
de saaie, grauwe achtergrond van de moderne grootsheid, ongeveer zoals
de muren van de Sala dei giganti van Giambattista Bologna in het Palazzo del
Te in Mantua: ruimte die wordt vervormd door het overweldigende panorama

domesticity, but will do so with an appropriate sense of incongruence and as a result will provide none of its comforts.
The television Shine 5.0 proposes to install plays a subversive role – not without its ironies – and completes the spatial and conceptual inversion of their living room. Television is the prime suspect of the post-modern condition: real life follows media. Instead of using it the way television appears to work best, that is by promoting and disseminating the culture of hypersensitivity to the outside and by encouraging a passive response to it, the television will be used to offset the frightening cinematographic non-walls by neutralizing them through the projection of stories and monologues about personal and everyday experiences of local space.
It is difficult to see whether the Shine group will achieve their heroic aim of tweaking the urban consciousness into learning to live with the modern sublime; or indeed whether they will succeed in their brave aim of domesticating those spaces which have been infected by events and local myths.
I fear their living room will merely exacerbate the all-too human wish for blind erasure by becoming part of the landscape they are trying to improve, simply through the incongruence of the attempt. Perhaps that is not so sad. After all, they say themselves that fear is good. At least such spaces speak of doom in a reasonably clear language. The possibility of violence is always real and it is good to accommodate that possibility in rhetorically similar spaces. In any case the act of domestication is two-faced at best. Domestication is an ambivalent form of evolutionary progress because it systematically chooses the more fragile complement in every binary opposition: man/woman, wild/tame, rough/gentle, etcetera and achieves an extraordinary docility of which the most resonant example, certainly within a Dutch context, is the tragedy of the large-uddered cow. The cow has sacrificed much to achieve its haven, even, paradoxically, the hope of a peaceful old age. The domestication of savage places will bear a monstrous fruit, perhaps even indifference.

Jacob Voorthuis teaches architectural critique at the Eindhoven University of Technology. Between 1994 and 1999, he was head of the humanities department at the Caribbean School of Architecture in Kingston, Jamaica.

van de vernietiging. De non-muren van de nagebouwde woonkamer zijn
complete cinematografische fresco's van een sombere achtergrond, waarop
de bezoeker op zijn gemak zijn eigen persoonlijke portretten kan projecteren
van het mogelijke. De muren zijn virtuele afscheidingen, niet fysiek maar
cinematografisch. Ze definiëren een voorwaardelijke ruimte, een ruimte
vol oncontroleerbare mogelijkheden. De woonkamer als nabootsing van de
veilige thuishaven, met de voorwerpen van een vredige huiselijkheid, maar
met een gepast gevoel van ongerijmdheid, zodat hij niets van het bijbehorende comfort te bieden heeft.
De televisie die Shine ook wil installeren, speelt een subversieve rol die niet
zonder ironie is en de ruimtelijke en conceptuele omkering van hun woonkamer completeert.Televisie is de hoofdverdachte van de postmoderne conditie: het echte leven volgt de media. In plaats van de televisie te gebruiken
zoals ze het best lijkt te werken, namelijk voor het promoten en verspreiden
van de cultuur van overgevoeligheid voor de buitenwereld en het aanmoe-
digen van een passieve reactie daarop, gebruikt Shine haar als tegenwicht
voor de angstaanjagende cinematografische non-muren, door ze te neu-
traliseren via de projectie van verhalen en monologen over persoonlijke en
alledaagse ervaringen van de omgeving.
Het valt moeilijk te voorspellen of de Shine-groep zijn heroïsche doelstelling
zal bereiken, namelijk het stedelijk bewustzijn zo fijn afstemmen dat we leren
leven met de moderne grootsheid; of ze überhaupt zullen slagen in hun moe-
dige opzet deze ruimtes te domesticeren, besmet als ze zijn door gebeurtenissen en lokale mythen. Ik ben bang dat hun woonkamer het al te menselijke
verlangen naar blinde vernietiging alleen maar zal verergeren en dat hij onderdeel zal worden van het landschap dat ze proberen te verbeteren, simpelweg
omdat hun poging zo ongerijmd is. Misschien is dat niet zo erg. Tenslotte
zeggen ze zelf dat angst goed is. Het doemdenken wordt door ruimtes als
de hunne tenminste in redelijk duidelijke taal verwoord. De mogelijkheid van
geweld is altijd reëel en het is goed die mogelijkheid in retorisch gelijksoortige
ruimtes een plaats te geven. In elk geval is de daad van domesticering op zijn
best dubbelzinnig. Domesticering is een ambivalente vorm van evolutionaire
vooruitgang, omdat ze bij elke binaire tegenstelling systematisch kiest voor
de fragiele component: man/vrouw, wild/tam, ruw/zachtaardig, enzovoort,
en leidt tot een uitzonderlijke onderworpenheid, waarvan het meest spreken-
de voorbeeld, zeker binnen de Nederlands context, wordt gevormd door de
tragedie van de koe met de grote uier. De koe heeft veel moeten opofferen
om zich van een veilig onderkomen te verzekeren, zelfs, paradoxaal genoeg,
de hoop op een vredige oude dag. De domesticering van wilde plaatsen zal
een monsterlijke vrucht dragen, misschien zelfs onverschilligheid.

*Jacob Voorthuis is docent architectuurkritiek aan de Technische Universiteit
Eindhoven. Van 1994 tot 1999 was hij hoofd van de humanities department
aan de Caribbean School of Architecture in Kingston, Jamaica.*

ANGST &RUIMTE

DE VISIE VAN JONGE ONTWERPERS IN NEDERLAND

FEAR &SPACE

THE VIEW OF YOUNG DESIGNERS IN THE NETHERLANDS

NAi UITGEVERS/PUBLISHERS

CONTENTS

INHOUDSOPGAVE

PREFACE
Urban Affairs (Theo Hauben & Marco Vermeulen)

Western society is terrorized by fear. Whether it be national security, road safety or public health; every form of risk has to be spread, covered or eliminated. This obsession with safety and security is leaving increasingly visible traces, and that in the public space. Speed ramps, smoke-free zones and surveillance cameras serve a society that runs on prevention. The mechanisms for control and exclusion by which we defend ourselves from the imagined disaster establish a norm and are thereby reductive. The desire for security has cut the environment we live in down to the bone and standardized it. The sharp edges have literally been knocked off. Our security Utopia prescribes a sterile, transparent and neutral environment where being normal is the norm and the exceptional is experienced as a threat. The issue of security can also be approached from the other side: what risks are we actually prepared to take? How does the environment we live in look if we ourselves are able to determine, or at least interpret, where the boundaries lie? In which structuring manner can security and risk give shape to the organization of the Netherlands?

Group Portraits
The obsession with security is the subject of *GP04, Group Portraits of Young Architects 2004*. This is a project set up by the Netherlands Architecture Fund and the Netherlands Foundation for Visual Arts, Design and Architecture and is being organized by Urban Affairs. *GP04* is the fourth in a series of projects which is trying to make use of the enthusiasm and open-mindedness of young designers for matters that have a great influence on the spatial organization of the Netherlands. *Group Portraits* lies on the boundary between research and design. Talented young designers are invited to give their views on a topical subject together with colleagues and contemporaries. In the course of the project the teams are fuelled by a collective programme of presentations, discussions and talks.

Presentation
This publication, which deals with the projects by the four design teams, is appearing at the same time as the closing *GP04* event in Rotterdam, where the results were also taking shape in the form of spatial installations (DUS), an event (Shine 5.0), a VJ evening (Mr. Smith), an exhibition and a debate (Untitled).
In this publication the projects are on the one hand explained by the teams and on the other put into perspective by writers who were not involved. The way this is done is different for each of the four projects. Benjamin Barber and Joshua Karant give a diagnosis of what is presented to them as Mr. Smith's syndrome.
The work done by Shine 5.0 is approached by Jacob Voorthuis from the point of view of architectural history and the etymology of architectural vocabulary.
Mark Pimlott puts the theme into an historical perspective and puts the design attitude displayed by DUS in its place on the timeline. Lastly, Moritz Küng curates the project by Untitled and leads it into the discourse on art.
Bert de Muynck has followed the teams throughout and introduces the projects by way of his personal view of fear and space.

VOORWOORD
Urban Affairs (Theo Hauben & Marco Vermeulen)

De westerse samenleving wordt geterroriseerd door angst. Of het nu de nationale veiligheid betreft, de verkeersveiligheid of die van de volksgezondheid – iedere vorm van risico dient te worden gespreid, afgedekt of uitgeroeid. De veiligheidsobsessie laat steeds meer zichtbare sporen achter, met name in de openbare ruimte. Snelheidsbeperkende drempels, rookvrije zones en surveillancecamera's dienen een maatschappij in het teken van preventie. De mechanismen van controle en uitsluiting waarmee we ons tegen het vermeende onheil verdedigen, werken normatief en daarmee reductief. Het verlangen naar veiligheid heeft de leefomgeving uitgekleed en gestandaardiseerd. Letterlijk zijn de scherpe kantjes ervan af. De veiligheidsutopie schrijft een steriele, transparante en neutrale omgeving voor, waar normaal zijn de norm is en waarin het bijzondere als bedreiging wordt ervaren. Het veiligheidsvraagstuk kan ook van de andere kant worden benaderd: welke risico's zijn we eigenlijk bereid te nemen? Hoe ziet onze leefomgeving eruit als we zelf mogen bepalen, of op zijn minst interpreteren, waar de grenzen liggen? Op welke structurerende wijze kunnen veiligheid en risico gestalte geven aan de inrichting van Nederland?

Groepsportretten

De veiligheidsobsessie is onderwerp van *Groepsportretten van jonge architecten 2004: GP04*. *Groepsportretten van jonge architecten* is een project van het Stimuleringsfonds voor Architectuur en het Fonds voor Beeldende Kunsten Vormgeving en Bouwkunst en wordt dit jaar georganiseerd door Urban Affairs. *GP04* is de vierde aflevering van deze projectenreeks, die tracht het enthousiasme en de onbevangenheid van jonge ontwerpers aan te wenden voor vraagstukken die van grote invloed zijn op de ruimtelijke inrichting van Nederland. *Groepsportretten* is gepositioneerd in het grensgebied van onderzoek en ontwerp. Talentvolle jonge ontwerpers worden uitgenodigd om tezamen met vakgenoten en generatiegenoten uit andere disciplines hun visie te geven op een actueel onderwerp. Tijdens het project worden de teams gevoed door een collectief programma van tussentijdse presentaties, discussies en lezingen.

Presentatie

Deze publicatie, die de projecten van de vier ontwerpteams bevat, verschijnt tegelijkertijd met de slotmanifestatie van *GP04* in Rotterdam, waar de resultaten tevens gestalte krijgen als ruimtelijke installaties (DUS), een event (Shine 5.0), een vj-avond (Mr. Smith), een tentoonstelling en een debat (Untitled). In deze publicatie worden de projecten toegelicht door de teams en tegelijkertijd in perspectief gezet door externe auteurs. De wijze waarop dit gebeurt, is voor elk van de vier projecten verschillend. Benjamin Barber en Joshua Karant geven een diagnose van het als zodanig aan hen gepresenteerde ziektebeeld van Mr. Smith. Het werk van Shine 5.0 wordt door Jacob Voorthuis benaderd vanuit de architectuurgeschiedenis en de etymologie van het architectonisch vocabulaire. Mark Pimlott zet het thema in een historisch perspectief en plaatst de ontwerpattitude van DUS op de tijdslijn. Moritz Küng ten slotte is curator van het project van Untitled en introduceert het in het discours van de beeldende kunst. Bert de Muynck heeft de teams gedurende het project gevolgd en leidt de projecten in met zijn persoonlijke visie op angst en ruimte.

THE PROSTHETIC PARADOX
Bert de Muynck

There can be no doubt that within the contemporary Western condition, fear is the driving force behind the (re-)organization of the public and private space. Fear is a factor that, due to its destabilizing influence, is expelled and excluded from society. Yet the way in which we do this makes a meaningful discussion about the relationship between fear and space impossible. The issue of fear and space is currently approached in such a way that the solutions are becoming invisible, in size and in function. This leads to an age of permanent justification, control, recording and interpretation. Technological prostheses installed in strategic locations keep our collective and individual fear under control. Simultaneously, these are signals that indicate that thinking about architecture and urban design has abandoned society. Architecture and urban design, both once specific organizational answers (as controlling as they were liberating) to societal debates, have divested themselves, from the second half of the twentieth century, from the idea that they could be, as form and programme, a means of social progress and/or reflection. They have now become an economic, political, cultural or social end – employable anywhere, simultaneously and with equal value. The current societal debate still relates to a significant extent to the discourse on progress, but architecture and urban design are no longer fundamental participants. The city and architecture are no longer a theatre of progress; instead they have literally become a decor. This is what some would have us believe. There is a form of truth in it, but it is behind this decor that an odd shadow play is taking place. On and behind the decor we see black contours, silhouettes and shadows moving, inspiring anxiety because they are unknown. What can we do? Watch anxiously or tear down the decor?

In *Eurotaoïsme* (1989) Peter Sloterdijk draws the following conclusion on the modern era: 'There is only one period called the modern era, because in the new era the capacity of Western people to act fascinated them to such an extent that they found the courage to proclaim the arrangement of the world purely through their own actions.'[1] For Sloterdijk the project of the modern era is based on a 'kinetic Utopia': every movement in the world should bring about our design of it: 'progress is movement into movement, movement into multiple movement, movement toward increased capacity for movement.'[2] The modern architectural and urban project has been characterized since 1950 by a shift from inclusive to exclusive, from synthesis to

DE PROTHETISCHE PARADOX
Bert de Muynck

Het leidt geen twijfel dat binnen de hedendaagse westerse conditie angst dé drijvende kracht is voor de (her)organisatie van de publieke en private ruimte. Angst is een factor die omwille van zijn destabiliserende invloed uit de maatschappij wordt verdreven en uitgesloten. Maar de wijze waarop we dit doen, maakt een zinvol debat over de relatie tussen angst en ruimte onmogelijk. Het vraagstuk van angst en ruimte wordt momenteel zo benaderd dat oplossingen in grootte en functie onzichtbaar worden; op strategische locaties aangebrachte technologische prothesen houden onze collectieve en individuele angst onder controle. Tegelijkertijd zijn het signalen die er op wijzen dat het denken over architectuur en stedenbouw de maatschappij verlaten heeft.

Architectuur en stedenbouw, beide ooit specifiek organiserende antwoorden (zowel bevrijdend als controlerend) op maatschappelijke debatten, ontdeden zich sinds de tweede helft van de twintigste eeuw van het idee dat ze als vorm en programma een middel tot maatschappelijke vooruitgang en/of reflectie konden zijn. Nu zijn ze verworden tot een economisch, politiek, cultuur of sociaal maatschappelijk doel; overal, gelijktijdig en gelijkaardig inzetbaar. Het hedendaagse maatschappelijke debat heeft nog steeds in belangrijke mate betrekking op het discours van de vooruitgang, maar architectuur en stedenbouw nemen hier niet meer fundamenteel aan deel. De stad en de architectuur zijn geen theater van vooruitgang meer, maar zijn letterlijk een decor geworden. Dat willen sommigen ons doen geloven. Er zit een vorm van waarheid in, maar het is achter dit decor dat een vreemd schimmenspel wordt opgevoerd. We zien op en achter het decor zwarte contouren, silhouetten en schaduwen bewegen die door hun onbekendheid angst inboezemen. Wat kunnen we doen? Angstig toekijken of het decor slopen?

In *Eurotaoïsme* (1989) stelt Peter Sloterdijk met betrekking tot de moderne tijd het volgende: 'Er bestaat alleen maar een periode die moderne tijd heet, omdat in de nieuwe tijd het vermogen van westerse mensen om te handelen zo'n fascinerende indruk op hen heeft kunnen maken dat ze de moed vonden de inrichting van de wereld louter door eigen handelen te proclameren.'[1] Voor Sloterdijk berust het project van de moderne tijd op een 'kinetische utopie': alle bewegingen van de wereld moeten ons ontwerp ervan realiseren: 'vooruitgang is beweging ter beweging, beweging ter meerdere beweging, beweging naar toegenomen bewegingsvermo-

opposition and from freedom to control. Anxiety has replaced fascination. We camouflage this by making our actions and arrangements dependent on prosthetics which we view as logical components of the kinetic Utopia. If architecture and urban design want to make a meaningful contribution to the movement, they must escape this logic. Otherwise we are headed – the signs are there already – toward a petrified city.

Collective Versus Individual
Until a few decades ago, designers strived for interaction with the modern project – as a total (Le Corbusier, among others), structural (the Smithsons, Maki, among others) or reactive project (Archigram, among others). In this evolution the demonstration of a shape of and for the modern era was the driving force. The failure of the partial execution of this led to the analysis that total transformation was not a good thing. The reasoning was that if the synthesis of architecture and urban design could not formulate an answer to the collective needs of the modern era, then its ambitions had to be broken down until it provided an answer to individual needs. City and architecture no longer relate in design to the collective, but to the individual. Sprawl as an ideal. The individual problem of Western man (including anxiety, security, culture, consumption, speed) became the global problem of the city and its architec-ture. In order to address this, we use prostheses – objects distributed here and there, both in the public and private space, which record, process and issue signals. The ad hoc and improvised way in which these have been distributed has led to a constant sense of dissatisfaction within society; waiting until problems appear before one resolves them is seldom a successful strategy – certainly in architecture and urban design.
The conflict between architecture and urban design formulates an answer to individual problems in enclaves or gated architecture. This answer represents a fear of the city, its antithesis, which is composed of various architectures. We bring this dialectic to its synthesis in a culture of prosthetics; the prosthesis replaces a part that has been lost; it supplements, adds something to what remains and refines it. The reasons why the public space leads both to freedom and to fear (including anonymity, mass, chance) are provided with a functional answer (including identifiability, individual, rules). In the public and private space the prosthesis works in an identical way;

it isolates or provides, but it also has another aim: in one sphere control is the necessary component of freedom, in the other freedom is a necessary component of control. The public and private spheres thus seem to exist in a perfect dialectic, the synthesis of which is our society. Yet this dialectic is not perfect; just like the public and private space, control and freedom are permeable and osmotic. Complete segregation is both Utopian and dystopian – the contemporary hybrid is an inevitable paradox. What is the significance of design in this condition?

The Prosthetic Programme
The expulsion of fear from our space has been manifested in different forms and organizations: in the past kept out by crude architectural and technological means (ramparts, forts and canons), now prevented by sophisticated architectural and technological means. Fear keeps chance at bay. Prosthetic elements have palliated fear in a world with increasing move-ment capacity including hoods, epaulettes, pacifiers, helmets and suits of armour, iron lungs, chastity belts, Ladyshave razors, head rests, piercings, Pichler's prototypes and 'Wohnzimmer'. Now the prostheses that are supposed to eliminate public and private fear are security cameras, alarm systems, TFT monitors, magnetic contacts, PI technology, volumetric motion detectors, glass breakage detectors, interior and exterior sirens, voice and digital telephone transmitters, active infrared barriers, iris scans, and so forth.

Anxiety?
Sigmund Freud says in Hemmung, Symptom und Angst (1926) that anxiety is first experienced at birth, a situation in which feelings of displeas-ure are not psychologically processed or dis-charged. Freud links anxiety to the threat of the loss of self, and so anxiety is narcissistic; in order to survive in this new condition the child has at its disposal expressions of anxiety, which it uses to attract the mother's attention. Anxiety is a signal of abandonment.
In the course of development into a social being, the danger becomes undefined and an impersonal entity. In his development the child conforms to social examples. This leads to an increasing and structured emancipation of the person; he or she acquires control, replacing external supervision, of his or her own attitude toward life. Anxiety is characterized not so much by disappearance as by expectation, which is 'an unknown, threatening crowded place'. When

gen'.[2] Het moderne architecturale en stedelijke project kenmerkt zich sinds 1950 door een verschuiving van inclusief naar exclusief, van synthese naar oppositie, van verwachting naar mogelijkheid en van vrijheid richting controle. Angst verving fascinatie. Dit camoufleren we door ons handelen en inrichten te laten sturen door prothesen die we als een logisch onderdeel van de kinetische utopie beschouwen. Willen architectuur en stedenbouw een zinvolle bijdrage aan de beweging leveren, dan moet die logica doorbroken worden. Anders stevenen we af, de voortekenen zijn er al, op een verstarde stad.

Collectief versus individu
Tot voor enkele decennia streefden ontwerpers naar interactie met het moderne maatschappelijke project; als totaalproject (o.a. Le Corbusier), structurerend project (o.a. Smithsons, Maki) of reactief project (o.a. Archigram). In die evolutie werd de tentoonspreiding van een gestalte van en voor de moderne tijd de drijfveer. Het falen van de gedeeltelijke uitvoering ervan leidde tot de analyse dat totale verandering niet goed was. Als de synthese van architectuur en stedenbouw geen antwoord kon formuleren op de collectieve noden van de moderne tijd, redeneerde men, dan moesten haar ambities gedemonteerd worden tot ze een antwoord gaven op de individuele noden. Stad en architectuur verhouden zich in ontwerp niet meer tot het collectief, maar tot het individu. *Sprawl* als ideaal. Het individuele probleem van de westerse mens (o.a. angst, veiligheid, cultuur, consumptie, snelheid) werd het globale probleem van de stad en haar architectuur. Om die weg te nemen, gebruiken we prothesen; her en der verspreide objecten, zowel in de publieke als in de private ruimte, die registreren, verwerken en signalen geven. De ad hoc en improvisatorische wijze waarop die werden verspreid, leidde tot een constant maatschappelijk onbevredigend gevoel; de problemen pas oplossen als ze zich voordoen, s zelden een geslaagde strategie, zeker niet in architectuur en stedenbouw.
De spanning tussen architectuur en stedenbouw formuleert in een architectuur van enclaves een antwoord op de individuele problemen. Dat antwoord is een angst voor de stad, de antithese, die uit diverse architecturen bestaat. Deze dialectiek brengen we tot synthese in een prothetische cultuur; de prothese vervangt een verloren gegaan deel, ze vult aan, voegt aan wat overblijft iets toe en verfijnt. De redenen waarom de publieke ruimte zowel leidt tot vrijheid als angst (o.a. anonimiteit, massa, toeval) wordt functioneel van antwoord voorzien (o.a. herken-

baarheid, individu, regels). In de publieke en private ruimte werkt de prothese op identieke wijze, ze isoleert of voorziet, maar heeft ook een ander doel, in de één is de controle de noodzakelijke component van vrijheid, in de ander de vrijheid een noodzakelijke component van controle. Zo lijken de publieke en private ruimte in een perfecte dialectiek te leven, waarvan onze maatschappij de synthese is. Maar die dialectiek is niet perfect, zoals de publieke en private ruimte zijn ook controle en vrijheid permeabel en osmotisch. Totale scheiding is utopie én dystopie, de hedendaagse tussenvorm is een noodzakelijke paradox. Wat betekent ontwerpen in deze conditie?

Het prothetische programma
Het verdrijven van angst uit de ruimte manifesteerde zich in verschillende vormen en organisaties: vroeger buitengesloten met ruwe architecturale en technologische middelen (omwallingen, forten en kanonnen), nu verhinderd door gesofisticeerde architecturale en technologische middelen. Angst houdt het toeval buiten. Prothetische elementen die persoonlijke angst verhielpen in een wereld met toegenomen bewegingsvermogen waren onder meer: capuchons, epauletten, fopspenen, helmen en harnassen, ijzeren longen, kuisheidsgordels, ladyshaves, neksteunen, piercings, Pichlers prototypen en 'Wohnzimmer'. Nu zijn de prothesen die de publieke en private angst moeten wegnemen beveiligingscamera's, alarmsystemen, TFT-monitoren, magneetcontacten, PI-technologieën, volumetrische bewegingsdetectoren, glasbreukdetectoren, binnen- en buitensirenes, vocale en digitale telefoonmelders, actief-infraroodbarrières, irisscans, enzovoort.

Angst?
Sigmund Freud stelt in *Hemmung, Symptom und Angst* (1926) dat angst bij de geboorte voor het eerst wordt doorleefd; een situatie waarin onlustgevoelens psychisch niet worden verwerkt of afgevoerd. Freud verbindt angst aan het gevaar van zelfverlies, en dus is angst narcistisch. Om in die nieuwe conditie te leven heeft het kind angstuitingen ter beschikking, die hij gebruikt om de aandacht van de moeder te trekken. Angst is een signaal van verlatenheid. In de ontwikkeling tot sociaal wezen wordt het gevaar onbepaald en een onpersoonlijke instantie. In zijn ontwikkeling sluit het kind aan op sociale voorbeelden. Dit leidt tot een toenemende en gethematiseerde emancipatie van de persoon, hij of zij krijgt controle, in plaats van

'the I' (or ego) in this situation is strong enough, it is ready to encounter forms of existence that are not yet part of the structure of its environment. Anxiety signals enable the ego to protect itself against danger before it gets the upper hand. Techniques to ward off danger, and therefore avoid anxiety, are defence mechanisms.

Anxiety signals mean a disruption of the integrity of the personality, which produces resistance. This is expressed as repression, 'isolation in relation to consciousness', 'exclusion from reproduction by memory' and anticipation. Another defence mechanism is the ego's reaction formation. To Freud, the most essential defence is regression, resulting in an 'introjection' and identification with what cannot be mastered. In the worst instance this regression does not occur and the personality can become fixated: 'fixated to such an extent that any form of openness to other forms of being and life are abandoned or rejected, so that the defence becomes an end in itself.'[3]

The Modern City

Space is the interaction of all the forms, directions, positions, distances and dimensions, conscious or not, which together with kinetic Utopia result in a collection of impressions, experiences, situations and perceptions to which we respond instinctively, emotionally, psychologically and physically. The connections and meanings among these are controlled by our thinking, instinct and character. Space has self-awareness and memory and develops over time. For instance, 9/11 not only influenced how terrorism had to be responded around the world, but also how New York had to be reorganized. Fear of the modern city led people to embrace the modern city in the first half of the twentieth century, but this turned into resistance in the second half – a resistance that offered no reactive alternative. The second wave of modernization, roughly from 1950 onward, constructed its imagery out of a negation of the first. Feelings of unreality, alienation, consternation, attraction and anxiety in the face the moving chaos represented by the metropolis were not to be provided with an answer but instead radically disempowered, forgetting that these are the foundation upon which the city constantly reinvents itself. This radical disempowerment became a permanent condition and was not resolved publicly, but rather in certain places and certain zones. Anxiety was no longer kept outside the city, but outside its buildings. Fear became the only thing connecting the

neurotic private spaces. And it roamed the public space. The city was no longer a womb of cultural possibilities, but a test tube for cultural control. An urban trend slowly emerged: thematic architecture would eliminate personal anxiety, thematic urban design would eliminate communal anxiety. But what architecture keeps out is at the mercy of the city, and what the city excludes is at the mercy of architecture. Movement leads to more capacity for movement, access, freedom of choice, contact, relationships, and it occurs in situations that subject movement to strict controls. Our anxiety grows in direct correlation with our increased knowledge of the world; it is avoided and combated everywhere and at all times. Wherever this is not the case, man is left to the mercy of the survival of the fittest. We live, work and spend time in kinetic dystopias that some see as the harbingers of capsular civilizations, others as 'ecologies of fear'. The debate on this seems only to feed negativism. A miasma of anxiety hangs round the kinetic Utopia, in which the confrontation with the unknown and with chance, the disappearance of familiar structures and the loss of familiar behavioural responses lead to isolation, suppression and exclusion of the (un)known and the suspect. Wherever the kinetic Utopia settles in monumental forms or organization (including mosques, peripheral expansions, airports, nuclear power plants, headquarters) it becomes a formalistic epicentre of anxiety and paralyzes any meaningful and reactive cultural debate. Anxiety has paralyzed the world into a permanent now-condition, with now-problems and now-fears – which have to be resolved now. Architecture and urban design, as organizers of space, seem unequal to the challenge; their slowness and impotence take greater and more mythical proportions every day. They have been supplanted by prosthetics, which are constant radars, on the lookout for danger, and emit smart, efficient signals, as in precision bombing. But against which fear are we actually protecting ourselves? Architecture and urban design are both victim and suspect in the current dilemma. Yet is there really a spatial connection between the arrangement of the world and the anxiety that brings it about in ourselves and others?

Public-Private Prostheses

Structure and change can be subsumed into space: among other things, space can isolate, offer resistance, exclude, secure, suppress, include, reject, petrify, absorb, digest and camouflage. . . Through control, anxiety

toezicht-van-buitenaf, op de eigen levens-
instelling. Angst kenmerkt zich dan niet zozeer
door verdwijning, eerder door verwachting, die
'een onbekende, bedreigende volte' is. Als 'het
Ik' in die situatie sterk genoeg is, is het in staat
bestaansvormen te leren kennen die niet of
nog niet in de structuur van zijn leefwereld zijn
opgenomen. Angstsignalen stellen het Ik in staat
zich tegen het gevaar te beveiligen, vóór dit de
overhand krijgt. De technieken om gevaar af te
wenden, en zo angst te vermijden, zijn afweer-
mechanismen.
Angstsignalen betekenen een verstoring van
de integriteit van het karakter waartegen verzet
komt. Dat uit zich als verdringing, 'isolatie
ten opzichte van het bewustzijn', 'uitgesloten
worden van reproductie door het geheugen' en
anticipatie. Een ander afweermechanisme is de
reactieve karaktervorming in het Ik. Voor Freud is
het meest wezenlijke afweermiddel de regressie,
resulterend in een 'introjectie' en identificatie
met wat niet bemachtigd kon worden. In het
slechtste geval vindt deze regressie niet plaats
en kan het karakter verstarren: '... dermate
verstarren, dat elke vorm van openstaan voor
andere vormen van zijn en leven verloren gaat
of afgewezen wordt; daarmee wordt de afweer
tot een doel in zichzelf'.[3]

De moderne stad
Ruimte is het samenspel van al dan niet bewust
georganiseerde vormen, richtingen, posities,
afstanden en grootes die samen met de kinetische utopie leiden tot een verzameling van
indrukken, ervaringen, situaties en gewaar-
wordingen waarop we instinctief, emotioneel,
psychologisch en fysiek reageren. De verbanden
en betekenissen daartussen worden beheerst
door ons denken, instinct en karakter. Ruimte
heeft zelfbewustzijn en geheugen, en ontwikkelt
zich in de tijd. Zo had 9/11 niet alleen invloed
op hoe er mondiaal op terreur moest worden
gereageerd, maar ook op hoe New York
gereorganiseerd moest worden.
De angst voor de moderne stad leidde in de
eerste helft van de twintigste eeuw tot een
overgave aan het concept van moderne stad,
maar uitte zich in de tweede helft in verzet
daartegen. Een verzet dat geen reactief alter-
natief aanbood. De tweede golf van moderni-
sering, grosso modo vanaf 1950, bouwde zijn
beeld op vanuit het negatief van de eerste.
Gevoelens van onwerkelijkheid, vervreemding,
consternatie, aantrekkingskracht en angst voor
de bewegende chaos die de grootstad voor-
stelde, moesten niet van antwoord worden
voorzien maar radicaal ontkracht. Daarbij vergat

men dat die gevoelens de basis zijn waarop de
stad zichzelf constant heruitvindt. Die radicale
ontkrachting werd een permanente conditie
en werd niet publiekelijk opgelost, maar op
bepaalde plaatsen en in bepaalde zones. Angst
werd niet meer buiten de stad gesloten, maar
uit haar gebouwen. Angst werd wat de neuro-
tische private ruimtes nog met elkaar verbond.
En zwierf door de publieke ruimte. De stad was
niet meer een baarmoeder van culturele mogelijkheden, maar werd een proefbuis voor culturele controle. Langzaam ontwikkelde zich een
stedelijke tendens: gethematiseerde architectuur
zou de persoonlijke angst wegnemen, gethematiseerde stedenbouw de gemeenschappelijke.
Maar wat de architectuur buitensluit, wordt
overgeleverd aan de stad, wat de stad uitsluit,
wordt overgeleverd aan de architectuur.
Beweging leidt tot meer bewegingsvermogen,
toegang, keuzevrijheid, contact, relaties en
gebeurt in situaties die de beweging aan
strenge controles onderwerpt. Evenredig aan
de toename van de kennis van de wereld groeit
onze angst; die wordt overal en te allen tijde
vermeden en bestreden. Waar dit niet gebeurt,
wordt de mens overgeleverd aan 'the survival
of the fittest'. We leven, werken en verblijven
in kinetische dystopieën die voor sommigen
voortekenen zijn van capsulaire beschavingen,
voor anderen 'ecologies of fear'. Het debat
daarover lijkt enkel de negativiteit te voeden.
Rond de kinetische utopie hangt een angstwaas
waarin de confrontatie met het onbekende en
het toeval, het verdwijnen van vertrouwde structuren en het verlies van vertrouwde gedrags-
reacties leiden tot isolatie, verdringing en uitsluiting van het (on)bekende en verdachte.
Waar de kinetische utopie in monumentale
vormen of organisatie neerslaat (o.a. moskeeën,
perifere uitbreidingen, vliegvelden, kerncentrales,
headquarters), wordt ze een formalistisch epicentrum van angst en verlamt ze een zinvol
en reactief beschavingsdebat. Angst heeft de
wereld verlamd tot een permanente nu-conditie,
met nu-problemen en nu-angsten. Die nu
moeten worden opgelost. Daartoe lijken architectuur en stedenbouw als organisatoren van
de ruimte niet in staat, hun traagheid en onvermogen nemen iedere dag grotere en mythischer
proporties aan. Prothesen hebben ze vervangen
en zijn constante radars, op zoek naar gevaar,
en geven, zoals bij precisiebombardementen,
slimme doelgerichte signalen af. Maar tegen
welke angst beschermen we ons eigenlijk?
Architectuur en stedenbouw zijn zowel slacht-
offer als verdachte in het huidige vraagstuk.
Is er wel een ruimtelijk verband tussen de

situations are no longer awaited or predicted, but prevented, warded off and isolated. In the best situation this leads to stability in the building, the street, the square or even the community itself, in the worst to a regression in which defence (against the mythical unknown) becomes an end in itself and certain sections of the population and certain lifestyles are denied access. They roam the public space and are moving anxiety signals; they confirm public anxiety because they are denied entry into the anxiety-free space.

New possibilities, technologies and structures, whether sociological, political, economic or architectural, have turned every space, since the beginning of the modern era, into a bewilderingly crowded place, described with concepts derived from Romanticism, like 'unheimlich', 'uncanny', 'Entfremdung', 'Entäusserung', 'alienation'. . . Later the public domain, the city as a free space, was further divided, sold off and compulsively arranged until it was fully controlled. A signal is sufficient to control public anxieties: orange for New York. Since 1950 large sections of the city have been turned over to private interests and turned into a slagheap of themed private initiatives. The only function of the government is to guarantee security on the slagheap. It attaches public prostheses to the surface of private vulgarity. The task of the government consists in guaranteeing security in a condition (the city) in which it has no say. The government is concerned with only one issue for the public space: security. The developer is concerned with only one issue for the private space: security.

The Prosthetic Utopia and Paradox

Yet the prosthetic Utopia provides a strange and unconvincing sense of comfort. There is a vague presumption that our need for security is being satisfied, but this is never confirmed, merely controlled. We see security failing, repeatedly. Where it succeeds, it exudes an artificial, dystopian temporariness. There is something out there. In this condition, fear becomes a means to achieve other objectives. Insecurity, the unexpected, freedom and chance are spatially delimited – no longer by a wall or glass façade, but by infrared. It seems that we can no longer create a future for the urban space, where anxiety is undoubtedly playing out, more than in the countryside. What then is the legitimacy to keep this modern project going?

There is a battle raging for public space, the mythological acme of anxiety-free liberty, which is being fought on two fronts, each with different means, each in different guises, but both with the same objective: in what remains of the public domain visible and hidden mechanisms of control have appeared, coupled with a masochistic mythologization of that public domain; the private domain is enclosed and equally controlled, leading to a narcissistic sense of security for the individual within the mass. Take the mass out of the latter, and all that remains, despite all precautionary measures, is fear. Take fear out of the former, and only the mass remains.

The spatial connection between the arrangement of the world and the anxiety that brings it about in ourselves and the other is the result of a prosthetic paradox that haunts our spaces: the public space is that of freedom and anxiety, the private space that of control and anxiety. The first wants more control in order to eliminate anxiety, the latter more freedom in order to eliminate anxiety. The first does this by deploying the prosthesis as an instrument of control, thereby making the space suspect, the latter by using the prosthesis as a free signal, making the person suspect. Increasing isolation, suppression and segregation of the public and private initiatives in the city will further petrify and thematize progress.

Contemporary anxiety is a signal of abandonment: of the city, of architecture, people, culture, the economy.

The only thing we will be able to proclaim in the future will be fear of our actions, fear of chance, fear of the shadow play in which we are the leading players. Unless the condition in which we live manages to fascinate us once again.

1   Peter Sloterdijk, *Eurotaoïsme*, Amsterdam 1991, p. 22.
2   Idem, p. 31.
3   Gustav Bally, *De psychoanalyse van Sigmund Freud*, Utrecht/Amsterdam 1965, p. 179.

Bert de Muynck is an architect-writer.
He publishes amongst others in Archis,
LDG, GAM and Archined.

inrichting van de wereld en de angst die
dit bij ons en de anderen teweegbrengt?

   Publiek-private protheses
In ruimte valt structuur en verandering aan
te brengen: ruimte kan onder meer isoleren,
weerstand bieden, uitsluiten, beveiligen,
verdringen, opnemen, afwijzen, verstarren,
absorberen, verteren en camoufleren...
Door controle worden angstsituaties niet
meer afgewacht of voorzien, maar verhinderd,
afgeweerd en geïsoleerd. In het beste geval
leidt dit tot stabiliteit in het gebouw, de straat,
het plein of de gemeenschap zelf, in het slechtste tot een regressie waar de afweer (voor de
mythisch onbekende) een doel in zichzelf
wordt, bepaalde bevolkingsgroepen en leef-
stijlen wordt de toegang ontzegd. Die zwerven
door de publieke ruimte en zijn bewegende
angstsignalen, ze bevestigen de publieke angst
doordat hun de toegang tot de angstvrije
ruimte wordt ontzegd.
De nieuwe mogelijkheden, technologieën en
structuren, zowel sociologisch, politiek, economisch als architecturaal, maakten met de
ingang van de moderne tijd iedere ruimte een
bevreemdende volte. Die omschreef met aan
de romantiek schatplichtige begrippen als
'unheimlich', 'uncanny', 'Entfremdung',
'Entäusserung' en 'alienation'. Later werd het
publieke domein, de stad als vrije ruimte, verder
gesplitst, verkocht en dwangmatig geordend
tot men er controle over kreeg. Om publieke
angsten te controleren, volstaat een signaal:
oranje voor New York. Sinds 1950 zijn grote
delen van de stad uit handen gegeven aan
privé-belangen en verworden tot een storthoop
van particuliere gethematiseerde initiatieven. De
enige taak die de overheid heeft, is de veiligheid
op de storthoop te garanderen. Ze kleeft publieke prothesen aan het oppervlak van de private
protserigheid. De taak van de overheid bestaat
uit de garantie van veiligheid in een conditie
(de stad) waarin ze geen inspraak heeft. Voor de
publieke ruimte heeft de overheid slechts één
thema: veiligheid. Voor de private ruimte heeft
de ontwikkelaar slechts één thema: veiligheid.

   De prothetische utopie en paradox
Maar de prothetische utopie geeft een vreemd,
en niet-overtuigend, gevoel van comfort. Er is
een vaag vermoeden dat onze behoefte aan
veiligheid bevredigd wordt, maar dat wordt
nooit bevestigd, enkel gecontroleerd. We zien
de veiligheid falen, in herhaling. Waar ze slaagt,
ademt de ze de artificiële dystopische tijdelijk-
heid uit. *There is something out there.*

In die conditie wordt angst middel om andere
objectieven te bereiken. De onzekerheid, het
onverwachte, de vrijheid en het toeval worden
ruimtelijk gelimiteerd. Niet meer door een muur
of glaswand, maar infrarood. Zo lijkt het er
op dat we voor de stedelijke ruimte, waar zich
zonder twijfel de angst afspeelt, meer dan op het
platteland, geen toekomst weten te verzinnen.
Wat is dan de legitimatie om dit moderne project
nog in stand te houden?
Er woedt een strijd om de publieke ruimte, het
mythologische summum van angstvrije vrijheid,
die op twee fronten wordt beslecht, telkens met
andere middelen, telkens in andere gedaanten,
maar met hetzelfde doel: in wat nog rest van het
publiek domein is een opzichtige en verborgen
controle ontstaan, en tegelijkertijd wordt dat
publieke domein op een masochistische wijze
gemythologiseerd; het private domein is afge-
sloten en wordt eveneens gecontroleerd, wat
leidt tot een narcistisch veiligheidsgevoel van het
individu in de massa. Haal je de massa weg uit
die laatste, dan blijft er, ondanks alle voorzorgs-
maatregelen, enkel angst over. Haal je de angst
uit de eerste weg, dan rest alleen de massa.
Het ruimtelijk verband tussen de inrichting
van de wereld en de angst die dit bij ons en
de ander teweegbrengt, is het gevolg van een
prothetische paradox die door onze ruimtes
waart: de publieke ruimte is die van vrijheid
en angst, de private die van controle en angst.
De eerste wil meer controle om te angst weg
te nemen, de laatste meer vrijheid om de angst
weg te nemen. De eerste doet dat door de
prothese als controle in te zetten, dus wordt de
ruimte verdacht, de laatste door de prothese als
vrij signaal in te schakelen, waardoor de mens
verdacht wordt. Een toenemende isolatie,
verdringing en scheiding van de publieke en
private initiatieven in de stad zullen de vooruitgang verder verstarren en thematiseren. De
hedendaagse angst is een signaal van verlatenheid: van de stad, de architectuur, de mens,
de cultuur, de economie. Wat we in de toekomst
nog kunnen proclameren, is de angst voor ons
handelen, angst voor het toeval, angst voor het
schimmenspel waarin we zelf de hoofdrolspelers
zijn. Tenzij we opnieuw gefascineerd kunnen
worden door de conditie waarin we leven.

1 Peter Sloterdijk, *Eurotaoïsme*, Amsterdam 1991, p. 22.
2 Idem, p. 31.
3 Gustav Bally, *De psychoanalyse van Sigmund Freud*,
   Utrecht/Amsterdam 1965, p. 179.

Bert de Muynck is architect-schrijver.
Hij publiceert onder andere in Archis,
LDG, GAM en Archined.

MR. SMITH

Mr. Smith could be anyone, but in the case of *GP04* he represents the collective identity of five individuals ranging in discipline from architect to researcher, graphic designer to video jockey. His five alter-egos are Mark van Beest, Duzan Doepel, Claudia Linders, Minke Themans and Ronald Wall, with a resulting schizoid personality that permits him to view the world through particular filters.

His obsession for understanding the world in its totality necessitates this combination of mental constructs, thereby offering him the tools to analyse, dissect and re-construct his reality. Time, scale and the effects that global processes have on local perception and spatial patterns of occupation, form his basic fascinations. These fuel his desire to generate abstract three-dimensional reconstructions of this complex system, with the ambition to re-direct and broaden the current discussion regarding the public domain

Mark van Beest
the Netherlands, 1970
graphic/audiovisual design (Nog Harder),
Rotterdam

Duzan Doepel
South Africa 1971
architect / ADD, Rotterdam

Claudia Linders
the Netherlands, 1966
architect / Labeled, Amsterdam

Minke Themans
the Netherlands 1972,
visual communication / deMink, Rotterdam

Ronald Wall
Zimbabwe, 1966
architect-economic geographer / WALL,
Rotterdam

MR. SMITH

Mr. Smith kan iedereen zijn, maar in het geval van *GP04* vertegenwoordigt hij de gemeenschappelijke identiteit van vijf individuen uit uiteenlopende disciplines, van architect tot onderzoeker, van grafisch ontwerper tot videojockey. Zijn vijf alter ego's zijn Mark van Beest, Duzan Doepel, Claudia Linders, Minke Themans en Ronald Wall, met als resultaat een schizoïde persoonlijkheid die de wereld door afzonderlijke filters kan bekijken.

Zijn obsessie met het begrijpen van de wereld als geheel maakt deze combinatie van denkbeelden noodzakelijk, waarbij hem de middelen worden aangereikt om zijn realiteit te analyseren, te ontleden en te reconstrueren. Tijd, schaal en invloeden die mondiale processen hebben op plaatselijke waarneming en ruimtelijke bebouwingspatronen vormen zijn belangrijkste fascinaties. Deze voeden zijn wens om abstracte driedimensionale reconstructies van dit complexe systeem voort te brengen, met de ambitie om de huidige discussie over het publieke domein een andere richting te geven en te verbreden.

Mark van Beest
Nederland, 1970
grafische/audiovisuele vormgeving (Nog Harder),
Rotterdam

Duzan Doepel
Zuid Afrika, 1971
architect / ADD, Rotterdam

Claudia Linders
Nederland, 1966
architect / Labeled, Amsterdam

Minke Themans
Nederland, 1972
visuele communicatie / deMink, Rotterdam

Ronald Wall
Zimbabwe, 1966
architect-economische geograaf / WALL,
Rotterdam

DUS

DUS is not afraid of creating places. Our aim is to 'set down' social relationships between people in order to create in individuals a critical consciousness of themselves and others. For this project, therefore, we decided to work with people who focus on interpersonal communications in a professional capacity.

In order to bring about social relationships through architecture, we will have to give up any copyright claims. We are open to the possibility that matters other than our own wishes may find their way into our work and thereby influence the result. Architecture as a gift in which others are challenged to participate. Change is celebrated.

This will create places within the city that are mutable and open to multiple interpretations. Designs that confront each person's imagination. In this way we create opportunities for communication between the private and the public domain, between individuals.

Arja Boon
the Netherlands, 1974
socio-cultural researcher and organizational expert / Johnson Property Group, Sydney

Hedwig Heinsman
the Netherlands, 1980
architect / DUS, Rotterdam

Diana Kuip
the Netherlands, 1980
freelance journalist-columnist, Amsterdam

Roel Spits
the Netherlands, 1978
educator / Paulusschool, Amsterdam

Hans Vermeulen
the Netherlands, 1977
architect / DUS, Rotterdam

Martine de Wit
the Netherlands, 1977
architect / DUS, Rotterdam

DUS

DUS is niet bang om plekken te maken. Het doel is om sociale relaties tussen mensen 'neer te zetten' om zo een kritisch bewust zijn van individuen met zichzelf en de ander te creëren. Daarom is voor dit project gekozen om met mensen samen te werken die zich beroepsmatig met de communicatie tussen mensen bezighouden.

Om sociale relaties door middel van architectuur tot stand te laten komen, zal het auteursrecht moeten worden opgegeven. We houden de mogelijkheid open dat andere zaken dan onze eigen wensen een plaats krijgen in ons werk en zo het resultaat beïnvloeden. Architectuur als een geschenk waarin anderen worden uitgedaagd er iets mee te doen. Verandering wordt toegejuicht.

Zo ontstaan plekken in de stad die muteerbaar en multi-interpreteerbaar zijn. Ontwerpen die een confrontatie aangaan met eenieders fantasie. Zo openen we de mogelijkheden voor communicatie tussen het privé- en het publieke domein, tussen individuen.

Arja Boon
Nederland, 1974
sociaal-cultureel wetenschapper & organisatiedeskundige / Johnson Property Group, Sydney

Hedwig Heinsman
Nederland, 1980
architect / DUS, Rotterdam

Diana Kuip
Nederland 1980
freelance journalist-columnist, Amsterdam

Roel Spits
Nederland, 1978
onderwijzer / Paulusschool, Amsterdam

Hans Vermeulen
Nederland, 1977
architect / DUS, Rotterdam

Martine de Wit
Nederland, 1977
architect / DUS, Rotterdam

UNTITLED

Untitled is a collaboration between Milica Topalovic, Bas Princen and Office Kersten Geers David Van Severen, founded with the intention to make architecture, motivated to tackle any given topic with an architectural project.

Untitled wants to test (at odds with the contemporary architectural discourse) whether and how architectural design can still be poignant when put in parallel with urban scale enquiries. It is dedicated to formulating methods for composing (architectonic) spaces and to doing research by making projects.

Untitled wants to focus again on the 'profession'. It wants to regain the ability of being critical within its own medium. It practices architecture as a discipline of space and phenomenology, and tries to investigate the consequences of this idea.

Kersten Geers
Belgium, 1975
architect / Office Kersten Geers David Van Severen, Rotterdam

Bas Princen
the Netherlands, 1975
designer-photographer of public space, Rotterdam

David Van Severen
Belgium, 1978
architect / Office Kersten Geers David Van Severen, Ghent

Milica Topalovic
Yugoslavia, 1971
architect, Rotterdam-Belgrade

UNTITLED

Untitled is een samenwerking tussen Milica Topalovic, Bas Princen and Office Kersten Geers David Van Severen. Het bureau is opgericht met als doel architectuur te maken en is gemotiveerd om elk thema aan te pakken met een architectuurproject.

Untitled wil onderzoeken, in tegenstelling tot het huidige architectuurdiscours, of en hoe het architectonisch ontwerp nog stimulerend kan zijn als het gelijkgesteld wordt met onderzoeken op stedelijke schaal. Het bureau houdt zich bezig met de ontwikkeling van methodes om (architectonische) ruimten te creëren en met onderzoek door middel van projecten.

Untitled wil zich weer op 'het vak' richten. Het wil het vermogen terugwinnen om kritisch te zijn binnen zijn eigen medium. Het praktiseert architectuur als een discipline van ruimte en als fenomeen, en het probeert de gevolgen van deze opvatting te onderzoeken.

Kersten Geers
België, 1975
architect / Office Kersten Geers David Van Severen, Rotterdam

Bas Princen
Nederland, 1975
ontwerper-fotograaf openbare ruimte, Rotterdam

David Van Severen
België, 1978
architect / Office Kersten Geers David Van Severen, Gent

Milica Topalovic
Joegoslavië, 1971
architect, Rotterdam-Belgrado

SHINE 5.0

Shine 5.0 is a collaborative venture between the Shine Project Group Foundation and Caroline O'Donnell.

Shine Project Group is an interdisciplinary group that focuses on linking theory and research in the field of urban culture in projects that can have an impact on everyday life, urban development and society in a broader perspective. As researchers we use our backgrounds in various disciplines (architecture and mapping, urban culture and theory, fashion and photography, graphic design and moving image) to investigate, analyse, record and map urban phenomena.

As a project group we undertake, stimulate and support projects in the urban domain (on a commercial as well as a non-profit basis). We do this by initiating projects ourselves, entering into collaborative ventures and establishing links to other initiatives. SPG has a core of individuals working together – Ade Aboaba (UK), Petra van Bennekum (NL), Wiebe de Ridder (NL) and Jasper Springeling (NL); within Shine 5.0 the group works with architect Caroline O'Donnell (UK).

Ade Aboaba
Great Britain, 1976
architect & community engagement consultant /
Shine Project Group / Urbanistix, Manchester

Petra van Bennekum
the Netherlands, 1975
photography & creative direction / Shine Project
Group, Rotterdam

Caroline O'Donnell
Ireland, 1974
architect, Rotterdam

Wiebe de Ridder
the Netherlands, 1977
marketing & communication, cultural theory /
Shine Project Group, Rotterdam

Jasper Springeling
the Netherlands, 1972
graphic design & moving image / Shine Project
Group, Rotterdam

SHINE 5.0

Shine 5.0 is een samenwerking tussen Stichting Shine Project Group en Caroline O'Donnell.

Shine Project Group is een interdisciplinaire groep die zich richt op het verbinden van theorie en onderzoek op het gebied van stedelijke cultuur in projecten die van invloed kunnen zijn op het dagelijks leven, stedelijke ontwikkeling en de samenleving in breder perspectief. Als onderzoekers gebruiken we onze achtergronden in verschillende disciplines (architectuur & mapping, stedelijke cultuur & theorie, mode & fotografie, grafisch ontwerp en bewegend beeld) om stedelijke fenomenen te onderzoeken, analyseren, registreren en in kaart te brengen.

Als projectgroep ondernemen, stimuleren en ondersteunen we projecten in het stedelijke veld (op commerciële en non-profitbasis). Dit doe we door zelf projecten te initiëren, samenwerkingen aan te gaan en links te maken met andere initiatieven. SPG heeft een kern van samenwerkende individuen: Ade Aboaba (UK), Petra van Bennekum (NL), Wiebe de Ridder (NL), Jasper Springeling (NL) en werkt binnen Shine 5.0 samen met architect Caroline O'Donell (UK).

Ade Aboaba
Groot-Brittannië, 1976
architect & community engagement consultant /
Shine Project Group / Urbanistix, Manchester

Petra van Bennekum
Nederland, 1975
fotografie & creative direction /
Shine Project Group, Rotterdam

Caroline O'Donnell
Ierland, 1974
architect, Rotterdam

Wiebe de Ridder
Nederland 1977
marketing & communicatie, cultuurtheorie /
Shine Project Group, Rotterdam

Jasper Springeling
Nederland, 1972
grafische vormgeving & bewegend beeld /
Shine Project Group, Rotterdam

ACKNOWLEDGEMENTS

This publication is the result of *Group Portraits of Young Architects 2004 (GP04)*, a project of the Netherlands Architecture Fund and the Netherlands Foundation of Visual Arts, Design and Architecture.

Compiled and edited by
Urban Affairs (Theo Hauben, Marco Vermeulen) with Véronique Patteeuw, NAi Publishers

Introductions
Bert de Muynck
Final editing Dutch
Els Brinkman
Final editing English
D'Laine Camp
Translation Dutch-English
Gregory Ball
Pierre Bouvier
Translation English-Dutch
Bookmakers

Design
Stout/Kramer (Marco Stout, Marjolein Delhaas)
Photography
Frank Hanswijk: cover illustration,
Petrovsky & Ramone: 121,127,128,129

Printing
Drukkerij Die Keure, Bruges
Paper
Eurobulk, 135gr

Projectleading
Véronique Patteeuw, NAi Publishers
Publisher
Simon Franke, NAi Publishers

© NAi Publishers, Rotterdam, 2004

Thanks to
Lex ter Braak, Anne Hoogewoning, Janny Rodermond, Nynke Siccama, Jan Jongert, Ana Dzokic, Florian Boer, Stefan Bendiks, Aglaée Degros, Hanneke van Wel, Lars van Es, Gert Anninga, Johannes Schele, Caro Baumann (Urban Affairs).
Benjamin Barber, Im Sik Cho, Natanya Doepel, Celine Jeanne, Henk Elffers, Joshua Karant, Bert v.d. Knaap, Next Architects, Erica Nijburg, Marijn Schenk, Uno Smith, Harm Tilman, Bas Themans, Suitbert Schmitt (Mr. Smith). Mark Pimlott, Christine Rusche, Christiaan Luth, Johannes Weisser, Joris van Tubergen, Groep 4/5 van de Paulusschool, Amsterdam, DSPS, zomercamp Deliplein, C. Steinweg Handelsveem B.V., Bouwcarrousel B.V. (DUS). Pier Vittorio Aureli, Kirsten Van den Berg, Dan Budick, Asli Cicek, Bert Gellynck, Paul Gerretsen, Tanja Elstgeest, Wonne Ickx, Dieter Lesage, Richard Venlet, Willem Jan Neutelings and Michiel Riedijk, Wendelien Van Oldenbourg, Studio Maarten Van Severen, etc. (Untitled). Koenen Air, Petrovsky & Ramone, Thijs (www.nepr.nl), Angela Kuperus, Kim Bakkers, Silvia, Ricky, Nannet, Fields and all others involved (SHINE 5.0).

Available in North, South and Central America through D.A.P./Distributed Art Publishers Inc 55 Sixth Avenue 2nd Floor, New York, NY 10013-1507, tel +1 212 627 1999, fax +1 212 627 9484, dap@dapinc.com

Available in the United Kingdom and Ireland through Art Data, 12 Bell Industrial Estate 50 Cunnington Street, London W4 5HB tel +44 208 747 1061, fax +44 208 742 2319 orders@artdata.co.uk

NAi Publishers is an internationally orientated publisher specialized in developing, producing and distributing books on architecture, visual arts and related disciplines.

www.naipublishers.nl
info@naipublishers.nl

Printed and bound in Belgium
ISBN 90-5662-422-9

COLOFON

Dit boek is het resultaat van *Groepsportretten van jonge architecten 2004* (*GP04*), een project van het Stimuleringsfonds voor Architectuur en het Fonds voor Beeldende Kunsten, Vormgeving en Bouwkunst.

Samenstelling en redactie
Urban Affairs (Theo Hauben, Marco Vermeulen) met Véronique Patteeuw, NAi Uitgevers

Inleidingen
Bert De Muynck
Eindredactie Nederlands
Els Brinkman
Eindredactie Engels
D'Laine Camp
Vertaling Nederlands-Engels
Gregory Ball
Pierre Bouvier
Vertaling Engels-Nederlands
Bookmakers

Ontwerp
Stout/Kramer (Marco Stout, Marjolein Delhaas)
Fotografie
Frank Hanswijk: cover illustratie,
Petrovsky & Ramone: 121, 127, 128, 129

Druk
Drukkerij Die Keure, Brugge
Papier
Eurobulk, 135gr

Projectleiding
Véronique Patteeuw, NAi Uitgevers
Uitgever
Simon Franke, NAi Uitgevers

© NAi Uitgevers, Rotterdam, 2004

Dank aan
Lex ter Braak, Anne Hoogewoning, Janny Rodermond, Nynke Siccama, Jan Jongert, Ana Dzokic, Florian Boer, Stefan Bendiks, Aglaée Degros, Hanneke van Wel, Lars van Es, Gert Anninga, Johannes Schele, Caro Baumann (Urban Affairs).
Benjamin Barber, Im Sik Cho, Natanya Doepel, Celine Jeanne, Henk Elffers, Joshua Karant, Bert v.d. Knaap, Next Architects, Erica Nijburg, Marijn Schenk, Uno Smith, Harm Tilman, Bas Themans, Suitbert Schmitt (Mr. Smith). Mark Pimlott, Christine Rusche, Christiaan Luth, Johannes Weisser, Joris van Tubergen,

Groep 4/5 van de Paulusschool, Amsterdam, DSPS, zomercamp Deliplein, C. Steinweg Handelsveem B.V., Bouwcarrousel B.V. (DUS). Pier Vittorio Aureli, Kirsten Van den Berg, Dan Budick, Asli Cicek, Bert Gellynck, Paul Gerretsen, Tanja Elstgeest, Wonne Ickx, Dieter Lesage, Richard Venlet, Willem Jan Neutelings en Michiel Riedijk, Wendelien Van Oldenbourg, Studio Maarten Van Severen, etc. (Untitled). Koenen Air, Petrovsky & Ramone, Thijs (www.nepr.nl), Angela Kuperus, Kim Bakkers, Silvia, Ricky, Nannet, Fields en alle andere betrokkenen (SHINE 5.0).

NAi Uitgevers is een internationaal georiënteerde uitgever, gespecialiseerd in het ontwikkelen, produceren en distribueren van boeken over architectuur, beeldende kunst en verwante disciplines.

www.naipublishers.nl
info@naipublishers.nl

ISBN 90-5662-422-9